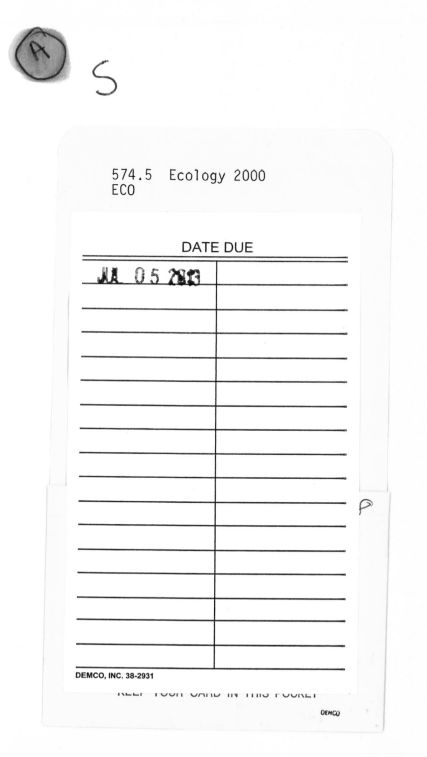

574.5 Ecology 2000
ECO

DATE DUE

JUL 05 2013	

DEMCO, INC. 38-2931

KEEP YOUR CARD IN THIS POCKET

DEMCO

ECOLOGY
2000
The changing face of Earth

Library of Congress Cataloging in Publication Data
Main entry under title:
Ecology 2000.
   Includes index.
   1. Eology. I. Hillary, Edmund, Sir.
QH541.E3195   1984   574.5   83-22376
ISBN 0-8253-0206-4

Published in the United States by Beaufort Books, Inc., New York.

This book was devised and produced by
Multimedia Publications (UK) Ltd

PRODUCTION DIRECTOR: Arnon Orbach
DESIGN AND ART DIRECTION: John Strange
PICTURE RESEARCHER: Sarah Waters
PICTURE ADVISOR: Tom Burke

Typeset by John Hills Typesetting, Sawbridgeworth, Herts.
Printed and bound by Sagdos SPA, Milan, Italy

Printed in Italy.   First American Edition

10   9   8   7   6   5   4   3   2   1

ECOLOGY 2000

The changing face of Earth
Edited by
SIR EDMUND HILLARY

Consultant **TOM BURKE**

Devised by **PETER NICHOLLS**

Beaufort Books, Inc.
New York

Learning About the Problems

I have been lucky enough over the years to be involved in a number of adventures – in the Himalayas, the Antarctic and elsewhere. But slowly my values changed – success on an adventure was still important but I had an increasing interest in human relationships. I became involved in assistance programmes in Nepal – building schools and hospitals, bridges and water pipelines. Success on a mountain was no longer the only thing that mattered; to help others to improve their way of life became a prime target. And so it has gone on – still the odd adventure . . . jet boats up the Ganges; through Tibet to the east face of Everest; backpacking on Baffin Island in the Canadian Arctic . . . but more and more I've been getting involved with people and their problems – and very satisfying it has proved to be.

As my interest in people has grown so too has my awareness of our natural environment and the importance of its conservation. After all, people and their environment are very closely related.

Thirty years ago conservation had not really been heard of. On our 1953 Everest expedition we just threw our empty tins and any trash into a heap on the rubble-covered ice at Base Camp. We cut huge quantities of the beautiful juniper shrub for our fires; and on the South Col at 26,000 feet we left a scattered pile of empty oxygen bottles, torn tents and the remnants of food containers.

The expeditions of today are not much better in this respect, with only a few exceptions. Mount Everest is littered with junk from the bottom to the top.

Since those years I have spent a great deal of time in the Himalayan Kingdom of Nepal. I have learned to understand the people, to enjoy their friendship and cheerfulness, and to gain an appreciation of some of their problems. One thing that has deeply concerned me has been the severe destruction that is taking place in their natural environment.

Population pressures are forcing the farmer higher and higher up the mountainside to find land where he can plant his crops. A large proportion of the forest cover has been destroyed in order to clear land for cultivation, to supply the local people with fuel and to produce firewood

Left: The scale of man's impact on his home planet is clear from this satellite photograph of the Libyan desert. The red dots are the vegetation from irrigation schemes, which are slowly making this desert at least bloom again.

Top: Sir Edmund Hillary

Above: Mount Everest

for trekking and climbing groups. The Nepalese are experts at ingenious and laborious terracing of their hillsides but when the monsoon rains come the surface soil is washed down into the streams, pours into the great Ganges river, flows out into the Bay of Bengal, and is finally deposited in the Indian Ocean. That valuable soil will never return.

The damage is not only being effected by the local people; foreign and international agencies are perhaps unwittingly causing their share as well. I have walked in from Kathmandu to the Mount Everest region perhaps thirty times and always enjoyed it, but a trip I made in March 1982 was different in many ways. We travelled by truck down the long valley to Dhologhat and crossed the big bridge built by Chinese engineers. Then it

was a slow climb over the ridge and down to the road beside the Sun Kosi river. We bumped our way up the valley, crossing many washouts and giant slips where the steep slopes were subsiding into the river. Then we reached the depressing shanty town of Lamsangu.

Now for the first time I crossed the new Swiss bridge over the Sun Kosi river and climbed slowly to the east up the new Swiss road. It was a fantastic route – very steep and difficult. We climbed higher and higher up over a steep pass then down again for thousands of feet, winding around sharp corners and very rough sections above steep drops. After seven hours of spectacular driving we reached our destination at Kirantichap and here we pitched our tents.

I found Kirantichap completely changed. It used to be a tiny bazaar in a dip in the ridge with several huge pipal trees shading it. It had always been a pleasant place to camp, although it had brisk winds at times. The trees were still there and so was the wind but there were now many houses and a big bazaar. The construction work on the road had produced a dusty desert and strips of slums. The whole mountainside ahead of

8

Hundreds of millions of tonnes of vital soil are washed off Nepal's hills each year. It is carried as silt down the Ganges where it often forms new islands. The island arrowed in the photograph is absent from maps made in 1971.

Roads are essential for development: they allow village products access to markets, and essential tools and machinery to be brought in. But they also encourage settlement on ecologically fragile land and do much direct damage to the environment through which they pass.

us up to Namdu and beyond – once so beautiful – was now terribly scarred from the work on the road. It was a depressing sight. The Swiss had gone to great trouble to build rock-retaining walls but on these steep loose hills the erosion would still be substantial. The engineering was superb but what possible use could the whole project have? Perhaps there were minor economic advantages for the area – no doubt food could be more easily transported out of Kathmandu – but what of the destruction of the mountainsides, the building of slums alongside the road, the devastating effect on the natural beauty of the area? Were the Swiss, I wondered, proud of what they were doing to this once stable and beautiful countryside?

Six days' walk further on we headed up the Solu valley taking the new high route above Beni. The whole beautiful Ringmo valley was scarred with a big new track, and the long hill down to Manedingma dropped steeply through wide stretches of destroyed forest. Why, I kept asking myself – the old road was nearly as short and very stable – why this obsession with destruction? There was only one reason. The United Nations were largely financing this work by donating imported grain and cooking oil. Without their support it would not have happened.

All the way up the Dudh Kosi river the track had been 'improved', with terrible scars and long stretches of eroded soil to show for it. Maybe the walking was a little easier and maybe the scars would heal again, but meanwhile vast quantities of soil would have washed out into the Indian Ocean and been lost for ever.

So I have become a keen and, I hope, a practical environmentalist. I am concerned not only about the deterioration of our environment in the affluent developed countries but in the poorer countries as well – those that simply do not have the finance to help themselves. I worry about the pollution in our great cities and in our many waterways. I even worry about the Antarctic and about the potential dangers facing that great remote continent.

I have spent much time in the Antarctic and I was last in McMurdo Sound in January 1982. I discovered that all the talk was about the oil potential and the possible mineral resources, and about farming the krill. Only the difficulties of access have prevented an even greater concentration on commercially oriented investigation and exploitation – seeking out some of the last miserable remnants of oil under the surface of the Earth. I heard little about the protection of this superbly beautiful environment, although much good work has been done by scientists in the past and is still being done. I dread the thought of drilling down through the moveable pack ice, with the possibilities of an enormous oil spill and the destruction of millions of Antarctic creatures.

The Antarctic Treaty has produced a demilitarized, unpolluted, wildlife sanctuary dedicated to free scientific co-operation. But now major political problems are looming. The possibility of economic development has turned the attention of many countries to the Antarctic, countries which have not signed

the Antarctic Treaty. Conservation could become a minor priority, in the search for wealth. When I was deeply involved in Antarctic exploration I regarded the South Pole as a continent of science and adventure; the world needs places like that and I hope it stays that way.

In June 1982, in London, I attended a Conference on the Human Environment organized by the United Nations Environment Programme. It was attended by distinguished scientists and administrators from all over the world and I found it a remarkable although rather terrifying experience. I am neither a distinguished scientist nor an administrator – I spent my early years as a simple bee farmer in New Zealand – but no one who has any feeling for the beauty of nature or any concern for his fellow humans could have been unaffected by what we heard.

Scientists told us the grim story of acid rain – of how in Scandinavia and Canada in particular thousands of lakes and great areas of soil had been poisoned by sulphurous fumes released into the atmosphere when fossil fuels were burnt.

I learned with concern of the carbon dioxide build-up in the atmosphere. There is still debate on the ultimate effects of this, but one thing is clear – the concentration of carbon dioxide in the atmosphere is increasing because of the burning of fossil fuels, deforestation and changes in land use. Dr Robert White, of the University Corporation for Atmospheric Research in the United States, forecast that in 60 years' time the amount of carbon dioxide in the atmosphere will have doubled and the average world temperature will have increased by four or five degrees. This, he believed, will melt sufficient Antarctic ice to raise the level of the ocean by fifteen feet, flooding many of our great cities.

There were even more depressing reports about situations with which I was more familiar. Desertification of the world's arable land threatens agriculture on every inhabited continent. Desertification, we were told, means any ecological change that saps a land of its ability to sustain agriculture or human habitation. It is perhaps seen at its worst in Africa. It can be controlled, the experts assured us, but only through substantial international effort: making changes in herding practices and land use, building fences, developing firewood alternatives, stabilizing sand dunes with hardy vegetation and so on. Effective action on a

Antarctica is the only continent which has not yet been significantly transformed by man. Its unique ecosystem is currently protected by the Antarctic Treaty, which comes up for review in 1990.

global basis, we were told, would require a commitment of several billion dollars a year from now to the end of the century. Inevitably it was pointed out that if only a tiny fraction of the world's armament bill was devoted instead to the environment then the world could bloom again.

One of the most devastating effects upon life on our Earth has been the enormous destruction of tropical rain forests, which is largely due to population growth but also aggravated by aggressive timber harvesting. The heavy concentration of the world's wild plant and animal species in tropical rain forests, I learned, meant that up to half of the world's genetic diversity is concentrated on six per cent of its land surface, and it is believed that one million species could be extinct by the end of the century.

A little time after the UNEP Conference I was encouraged to read a speech by Mrs Indira Gandhi, Prime Minister of India, calling for firm steps to save the environment in her great country. Opening the State Forest Ministers' Conference, she said that there is 'collective anxiety' in her country over the rapid depletion of forest resources. 'Yet', she told the Conference, 'when it comes to taking concrete decisions either to stop the cutting of trees or to preserve endangered species of animals or to put down poaching or smuggling of rare species, we waver.'

She wanted the avarice of the contractor to be recognized and dealt with firmly. Mrs Gandhi said that some hard measures needed to be taken, like a ban on the felling of trees in all the critically affected areas like hillslopes, catchment areas and tank-beds.

Many other topics were discussed during the Conference on the Human Environment, including the disposal of atomic waste, and even war itself. Each seemed more devastating than the last and it was hard to believe that humans could be so improvident and stupid.

After the first day of the Conference I walked rather sadly out of the Greater London Council Hall and into bright sunlight from a clear blue sky. With rising spirits I strolled across the green grass and superb trees of St James's Park. The world was still very beautiful, I told myself, whatever the future might hold. Every effort must be made to preserve it.

The second day of the Conference was largely devoted to discussing what practical action could be taken to alleviate these environmental problems. The International Conference on the Environment came to certain major conclusions: that the problems were severe and required urgent action and that time was desperately short; that the task could only effectively be tackled on a global scale and would require very substantial sums of money; and that governments would be reluctant to undertake such dramatic action unless they were pressured by strong and well-informed public opinion. It was felt that the NGOs (the Non-Government Organizations) had a very important role to play in the education of the public and in pressuring governments. People cannot push for action unless they know what the problems are. It was in response to the Conference's call for a clear presentation of information to ordinary people that this

The atmosphere has no frontiers — it is a 'resource' which is held in common by all mankind. There is growing evidence (see Chapters 7 and 10) that human activities are polluting the atmosphere on a global scale.

book, *Ecology 2000*, was first planned.

At the conclusion of the Conference the distinguished members were each asked a simple question: 'Are you optimistic or pessimistic about the future of our world?' Hardly one person in this international gathering was completely optimistic. All had varying doubts and concerns. But not one person was completely pessimistic either.

I came away from the Conference with the firm belief that the future is entirely in our own hands. We can make the world what we will, a paradise for all or a barren desolate globe spinning endlessly through space.

But the problems are certainly enormous. Regional and commercial interests exert tremendous pressure to mould government views, often with little interest in the long-term view and the welfare of future generations. Our only hope for the future must lie in a strong and well-informed public opinion and in those devoted people who work so energetically to protect our world from unnecessary exploitation and pollution. In the long run it is all up to us!

Most of us are unable to have a major impact on the world scene. Individually we can only try to deal with the challenges that arise in our own particular field, but even this can be well worth while. When I first visited Mount Everest in 1951, what a beautiful place it was. I can remember crossing the pass above Chaunrikarka and looking for the first time into the upper reaches of the Dudh Kosi river and seeing the sacred peak of Khumbila towering up in the heartland of the Sherpas. We climbed through dense pine forest up the long steep hill to Namche Bazar. The whole region was dense with greenery. Below the village, giant conifers soared, framing the snow and ice peaks that lined the other side of the valley. We climbed to Thyangboche Monastery at 13,000 feet; it was clothed in forest and surrounded by a ring of superb mountains.

We reached Pangboche village, with its ancient monastery and its tall gnarled juniper trees. Most of the junipers from here on were shrubs, but in places the forest remained, and there were ample supplies of firewood. When we turned into the Khumbu Glacier valley the forest had disappeared, but the dark green juniper bushes covered the slopes and yaks grazed on the dry grass. It was incredibly beautiful and dramatic.

About 25 years later I repeated this very same journey. The valley of the Dudh Kosi river was still very beautiful, but the forest was woefully thinned by the axes and saws of the Nepalese who had been cutting timber for buildings. The trees below Namche Bazar had been scarred by the heavy knives of Nepalese porters taking branches and bark for fuel and gummy heartwood for torches and lighting fires. The forests around Thyangboche had lost many of their mighty trees, and the Pangboche area was almost bare. Up the Khumbu Glacier valley there was hardly a juniper to be seen.

What had happened to produce such a change? Our climb of Mount Everest brought more mountaineers of many nations eager to attain the top of the world. Fuel for their expeditions rapidly exhausted supplies of the widespread juniper – but at first the forests themselves were left almost untouched.

In a way, I was initially responsible for the subsequent damage to the forests. In the early sixties I made an effort to assist my friends the Sherpas by building schools, hospitals, bridges and water pipelines. To help in the transport of building materials we constructed an airfield at Lukla. But the airfield had an unexpected effect: it gave much easier access to the Everest area, and increasing numbers of trekkers and tourists accelerated the demand for fuel.

The 1970s were a period of tremendous expansion in the Khumbu. Five thousand foreigners a year were now visiting the Everest region and for each foreigner it was estimated there were three Nepalese employees from outside

the area. Each year 20,000 people were coming into the Khumbu – a considerable burden on the 3,000 members of the local population. Dozens of small hotels were constructed. Tea shops and beer shops abounded. The weekly bazaar at Namche was thronged as hundreds of Nepalese vendors offered food and fuel to visitors and their porters. The forests suffered as the demand for firewood and building timber escalated.

By 1973 I was sure that some sort of control would have to be exerted if the Khumbu District were not to become a treeless desert. The Khumbu was such a remote area that Government administration was not only difficult but almost non-existent. Government funds were very hard to come by. The answer seemed to be the establishment of some form of National Park, initially financed by foreign aid.

In October 1973 I talked with the United Nations advisor on forestry in Kathmandu who had already been discussing the possibility of an Everest National Park with the Director of Nepal's small National Park Authority. It was felt that outside help was needed. New Zealand was a country with a topography similar to Nepal's and a well developed National Park system. Would New Zealand, they asked me, be prepared to help get the Sagarmatha (Everest) National Park established? I had no idea, but I approached the New Zealand Ministry of Foreign Affairs and, astonishingly, got immediate action. A three-man mission was sent to Nepal and

Above: improved communications, including the building of this airport, have brought economic benefits, but at a growing ecological price.

Top: today the forests that surrounded Namche Bazar 30 years ago have all gone, used for firewood and buildings. Without the protection of the cover, the rains wash the soil from the steep slopes.

reported favourably. In 1975 the first New Zealand National Park advisor moved up to Namche Bazar and the Sagarmatha National Park was under way.

An initial problem was the grave doubts that the Sherpas themselves had about the National Park. They worried that it might restrict their firewood supplies and limit their yak grazing pastures. There were even rumours that they might be moved out of the Park to leave it to the trees, wild animals and tourists, as had indeed happened in another Nepalese National Park. In the 1976 election of the local village councils, all the chief candidates were against a National Park. An old friend of mind, Khunjo Chumbi, was aware of my original support of the National Park and commented in his political speeches that 'Hillary first brought sugar to the lips of the Sherpas, but he is now throwing chili in their eyes'.

I had to agree that his concern was valid.

Initially the New Zealand wardens concentrated rather strongly on building Park headquarters and other structures, but slowly they came round to the view that the welfare and co-operation of the Sherpas were equally important, and a much happier balance was achieved. By the time the New Zealand five-year aid programme was completed some excellent work had been done and several forestry nurseries had been established. Sherpa Mingma Norbu, who spent five years training in New Zealand, was made chief warden of the Park and carried on the job very effectively.

With the withdrawal of New Zealand there was the major problem of how the forestry programme could be financed. In the end my Himalayan Trust agreed to contribute up to 15,000 dollars per year which Mingma Norbu felt was initially

The tropical rainforest, Terai, Nepal.

adequate for re-afforestation purposes. Major change is inevitably slow, particularly at altitudes above 12,000 feet, but improvement and change there have undoubtedly been. The birds and animals in the Park are increasing rapidly again. There are now many forest nurseries growing vigorously, and young trees are being planted out in extensive numbers. Five hundred goats, which had been recently introduced to the Park and were doing considerable damage to young growth, have been gathered up and removed from the area. It will take a long time for the Khumbu to become as it was thirty years ago but at least progress is being made and that is the best we can hope for.

Environmental problems are really social problems anyway. They begin with people as the cause, and end with people as victims. They are usually born of ignorance or apathy. It is people who create a bad environment – and a bad environment brings out the worst in people. Man and nature need each other, and by hurting one we wound the other. There is so much that needs to be done to halt the destruction of our world environment, so many prejudices and so much self-interest to be overcome. How can the situation possibly be changed in the time available?

My remaining hope is the amazing adaptability of human beings and the astonishing resilience of nature itself. Certainly the world and its human inhabitants are both changing, but we can hope that all the changes will not be bad. Perhaps humankind will start walking firmly in the direction of reconstruction and a better way of life. Maybe there is a good future for us all yet.

Sir Edmund Hillary

Chapter 2

Humanity's Growth

The conventional way to define 'ecology' is to say it is a science that deals with the relationships between living things and their environments. A more contemporary way to define the term might be to say that it relates to the unprecedented way in which a single species, *Homo sapiens,* has come to dominate the planetary ecosystem. True, some forms of ecology, the more esoteric forms, still interpret the science in terms of counting bugs on the back lawn, or some other micro-scale experiment. But a more realistic view is to consider the impact of humankind on all natural environments around the Earth – an impact that is being generated with the energy and the consequences of a major geologic upheaval, such as an Ice Age.

Three related factors contribute. The first is growth in human numbers. The second is growth in human technological capacity. The third is growth in human appetite for material goods. During the course of this chapter, we shall look at each of these factors in turn.

Our transformation of the face of the Earth began to gather steam about 50,000 years ago, when Man mastered the use of fire. It took a whole leap forward some 10,000 years ago, when Man learned the rudiments of agriculture. It started on a further giant stride about 200 years ago, with the onset of the Industrial Revolution. Its most recent and most marked advance has occurred in just the last three decades or so, as medical innovations have allowed a super-surge to the growth in human numbers.

Plainly this process, the process of Man's domination of the natural world around him, has been accelerating. Time was when humankind took tens of thousands of years to accomplish much change at all in its fortunes. Then the intervals became just a few thousands of years, then just a few hundreds. Finally we count the intervals in tens of years. In terms of our capacity to alter our planetary living space, we have probably accomplished as much in the years since 1950, as we did during the entire span between AD 0 and the year 1950. In just a few months now, we can expand deserts and cut down forests on a scale that would have taken decades in the time of our great-grandparents. A man with an axe would need one year to fell a forest the size of an average football field. The advent of the chainsaw has meant that the same man can do the job in ten days – although he would need much longer to clear away the felled timber. Today in Borneo, we can witness a machine that resembles a cross between a combine harvester and a giant tank. It not only cuts down every single tree, but it reduces each one to wood chips – boles, branches, roots, the lot. Within a single day, a couple of acres can lose virtually every last vestige of woody vegetation, perhaps 1,000 tonnes in all, from that football-field-sized space, perhaps two acres.

At the same time, there are many more people with many more demands for products that can be made from those wood chips – newspapers, magazines, boards, furniture, house walls, weekend yachts and so on. Everybody's hand is at the wheel of the chipping machine making its way across Borneo – 'everybody' now meaning very many people, growing still many more with every tick of the clock.

Growth in Human Numbers

I remember a professor at the University of California who was illustrating to his class the sudden upturn in human population growth during the recent past.

He proposed that we consider the blackboard, and reckon the long lower edge as a time scale, running from 500,000 years ago (roughly the time when Man became Man) to the present time; and that we view the vertical edge of the blackboard as a scale on which we could plot the totality of human numbers, from nought to five billion. The professor then drew a 'curve' to illustrate the history of human population increase. He started out at the bottom lefthand corner, with his chalk running pretty well along the bottom of the blackboard, until he reached almost the righthand edge, then, with a slight bend, he took the line into a virtually vertical plane up the righthand edge of the blackboard. He was emphasizing that the outburst in human numbers has been exceedingly recent, and exceedingly strong.

Let us take a closer look at the process. When Man adopted his earliest lifestyle, that of a hunter-gatherer, he probably occupied no more than about one third of the planet's land surface. Given what we know of the densities of hunter-gatherer tribes today, we can estimate that the total human population until 10,000 years ago could hardly have exceeded five million souls. The Agricultural Revolution permitted the first of many explosions of human numbers until, by the time of Christ, there were probably between 200 million and 300 million people. As agriculture expanded and became more sophisticated, so numbers increased yet more, until, by the time of the first stirrings of the Industrial Revolution around AD 1650, humankind probably amounted to about 500 million people.

Throughout the eighteenth century, the Industrial Revolution proceeded on its way with ever greater momentum. This brought about a further surge in human numbers, enabling humankind's total to top one billion (one thousand million) by the middle of the 1800s. After this momentous point in human history, a second billion was added far more rapidly,

17

Top: Benares, India. With more than 700 million people, India alone now has a larger population than the whole world prior to the Industrial Revolution.

Above: forest clearance in the Amazon. Well over a third of the Earth's original forest cover has already been cleared to create living space. This clearance continues at a rate of more than 121,000 hectares (300,000 acres) an hour.

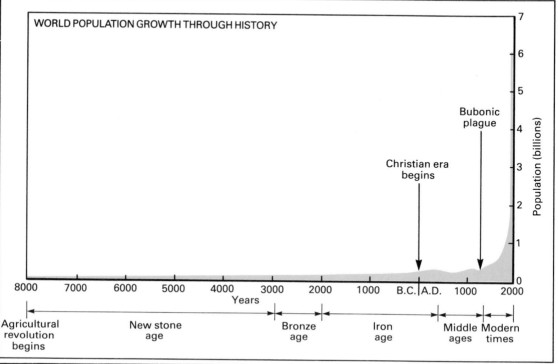

WORLD POPULATION GROWTH THROUGH HISTORY

Bubonic
plague

Christian era
begins

Population (billions)

8000 7000 6000 5000 4000 3000 2000 1000 B.C. A.D. 1000 2000
Years

Agricultural
revolution
begins

New stone
age

Bronze
age

Iron
age

Middle Modern
ages   times

Top: the Pattinum people of Brazil still live as
hunter-gatherers. Until agriculture began, about
10,000 years ago, hunter-gathering was the
dominant regime.

Above: the graph traces the growth in human numbers
over the past 10,000 years. In the last 200 years the
steady rise of population over millennia has been
replaced by the explosive growth we are experiencing
today.

probably by 1930. Immediately thereafter, because of the Depression and World War Two, there may have been little expansion to bother about. But from 1950 onwards, a new phenomenon emerged, preventive medicine in the tropics. Within just a few short years, vaccination campaigns put an end to the great pandemic diseases that used to kill off one child in five during the first year after birth, and another one out of five of the survivors by the time that schooling began. The result was that humankind reached four billion by 1975. As the reader looks through this book – perhaps early in 1984 – he or she can speculate that there are another 4.7 billion persons somewhere on the planet, so we shall not have to wait long to reach five billion – and it looks virtually certain that we shall exceed six billion by the year 2000.

An illuminating way to comprehend the process is to consider it in terms of doubling time – or the number of years needed for a population to double in size. During the period 8000 BC to AD 1650

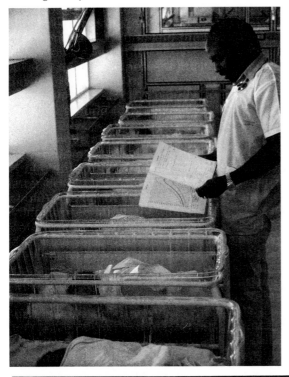

when the total expanded from about five million to 500 million, the doubling time averaged 1500 years. But the next doubling, from 500 million to one billion, required only 200 years; and the next doubling, from one to two billion, a mere 80 years. The latest doubling, to four billion by 1975, took only 45 years.

Again, let us look at the situation from a slightly different angle. For humankind to achieve a population size of one billion took half a million years. The next one billion was added during the period 1850–1930, only 80 years. The third billion arrived during the period 1931–1960, a trifling 30 years. The fourth billion came along virtually instantaneously, during the period 1961–1975, just fifteen years, way under a single generation.

Population Growth and the Third World
As is well known, the great bulk of population growth now occurs in Third World countries. In 1950, the global total of about 2.5 billion consisted of some 0.8 billion people in developed countries, and the rest in developing countries. Today, developed countries have expanded to only 1.2 billion, whereas developing countries have expanded to 3.4 billion. By the year 2000, developed countries are projected to expand hardly at all, to 1.3 billion, while developing countries are projected to increase to 4.8 billion. No doubt about it, the population explosion is a phenomenon of the developing world.

But Third Worlders have not suddenly started to breed like rabbits. Rather, they are no longer dying like flies. Their average family size, six to eight children, is no bigger than was often the case in the United States and Great Britain 100 years ago, when they too were, in effect, developing countries. But in those far-off days, when public health services were limited at best, many of the children failed to survive beyond the age of five. Today, by contrast, massive child mortality is, relatively speaking, a thing of the past. In Kenya, an average woman of

Improved medical technology, as in this maternity unit in Zimbabwe, has dramatically reduced infant mortality, but, in doing so, it has helped create new problems.

reproductive age produces almost nine children, virtually all of whom make it through the early critical years before school.

Still more to the point, the sudden vast improvement in child survival has meant that developing-world communities tend to have relatively more children than is the case with developed nations. Whereas persons aged fifteen and less in Great Britain amount to little more than one fifth of the populace, they constitute well over one half in many developing countries. This means that there is a vast amount of 'built-in' momentum for population growth to extend far into the future. Were all parents in, say, Kenya to reduce their family size to two children forthwith, Kenya's population, now eighteen million, would continue to grow for at least another two generations, and at least double before it reached zero growth rate. As it is, even with optimistic expectations for family planning campaigns, Kenya's population growth is

not projected to stabilize until the start of the twenty-second century, at a level of 109 million people.

From 1958 to 1982, I lived in Kenya. I remember the time when, just about at the point of Independence in late 1963, it was announced that Kenya's growth rate had apparently reached three per cent a year. Some people were concerned, and talked about 'runaway population growth'. How would Kenya feed all those new mouths, especially if the growth rate did not come down – and the Government showed few signs of wanting to intervene in the situation. Suppose the rate remained at three per cent for a good while to come: how many people would Kenya accumulate? There were some spirited guesses offered in the 'letters to the editor' columns of the newspapers. Some people suggested that the 1963 population of ten million might increase to 50 million within 100 years; others proposed that it could even reach 100 million within 100 years. Then one day an

Large families, like this one from Kenya, are as commonplace throughout the Third World as they once were in the industrialized countries.

experienced demographer wrote to the newspaper: he pointed out, with impeccable arithmetic, that a three per cent growth rate causes a population to expand 19 times within a century, so Kenya's total would soar to 190 million.

Today, Kenya's growth rate is at least four per cent a year. If this rate were maintained for 100 years, the population would increase fifty times over. Of course Kenya's growth rate is not going to remain at four per cent, since the Government is finally coming to grips with the situation, having learned that if people themselves do not limit their numbers, nature will. (Kenya has changed from a sizable food exporter to a large food importer in the last few years.) Nonetheless, there is so much 'inertia' in Kenya's population growth patterns that the *growth rate* of the growth rate will not itself be slowed for some years to come. If Kenya can find the funds to purchase sufficient food abroad, the growth rate might even rise to 4.5 per cent a year, before finally starting to decline. It would be an optimistic population expert who would predict that Kenya would have come back down to three per cent a year (unless massive starvation intervenes) by the year 2000. Thereafter of course we can hope that family planning programmes will start to 'bite' – far too late in the process.

So much for Kenya, and its dismal plight. A similar pattern applies to many other countries of black Africa, where growth rates are still rising – by contrast with Asia and Latin America, where the rates are falling, often quite rapidly (more details below).

How is the situation for the world as a whole? Well, far from facing a future where its numbers increase several times over (as in black Africa), humankind is projected to do little more than double its numbers before it levels out at zero growth, with a total around ten billion. Perhaps a little more, if things do not work out well, perhaps a lot less if more nations of the world could be persuaded to follow the example of the Chinese, who are making a norm out of the one-child family (the Chinese even appear to be aiming for a policy of an eventual *negative* growth rate).

But we can scarcely congratulate ourselves at the prospect of limiting

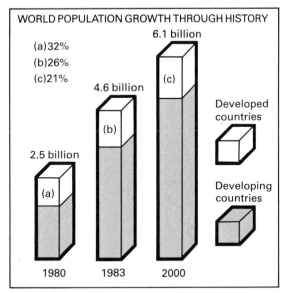

WORLD POPULATION GROWTH THROUGH HISTORY

(a) 32%
(b) 26%
(c) 21%

6.1 billion
4.6 billion
2.5 billion

1980   1983   2000

Developed countries

Developing countries

human numbers to ten billion. We have dreadful trouble in supporting the present 4.7 billion, in so far as 1.5 billion are inadequately nourished and half a billion are severely under-nourished (another half billion are distinctly over-nourished). True, there is enough food grown around the world to send every single global citizen to bed with a full stomach. But the food is unjustly distributed. However much Marxists may complain, human leaders are not likely to achieve the wisdom of saints overnight, and we must expect that the world will continue to witness the spectacle of starvation amid a system of plenty.

Regional Variations

So much for the overall picture. Let us note some variations at regional levels. Whereas the developed world has now brought its growth rate way below one per cent (Europe to only half of one per cent, and several countries even below replacement level), and Asia and Latin

The high birthrates in developing countries mean that they account for an ever larger share of the world's population. By the year 2000, the developed nations will account for less than a quarter of the planet's people.

America are now well below two per cent and plunging fairly fast, Africa is still stuck at around three per cent – and *rising*. Worse still, black Africa has become steadily hungrier since 1970; according to the World Bank and the United Nations, Africa's situation is not likely to improve this side of the year 2000.

What a bleak outlook – and what a declining prospect for Africa's natural environments. No person is more destructive of forests and brushlands than a hungry peasant (except, in some cases, a super-rich person – see *below*). During the years ahead, we can expect to see a rapid and far-reaching expansion of agriculture into all but the most inhospitable areas of the African continent.

Yet even this massive expansion of cultivated land is unlikely to avail much in the long run. Virtually all the fertile soils of Africa are already under crops. When forests are chopped down, the cleared land sustains agriculture for only a year or two before the soil's nutrients are used up, whereupon the farmer must move on

and chop down a further patch of forest. As for grassland zones and other savannahs, they are generally too dry to support conventional agriculture, and the top soil quickly blows away in the desiccating winds. Again, the peasant must constantly move on in search of new croplands. Unfortunately Africa is rapidly running out of new ground to break with a digging hoe.

One consequence along the way appears to be mounting pressure on the last holdouts of elephants, zebras and lions. No matter how committed the governments of Africa might be to the cause of wildlife conservation, the time may soon some when they could post armed guards at one-yard intervals around their parks and reserves, and still not keep out the landless peasants.

This is all written against my background as a resident of Kenya for almost one quarter of a century. When I finally took my leave of that lovely land in 1982, I found it immensely hard to tear myself away from the safari territories

Kanimboman, Chad — poor land, primitive tools and fast-rising population are the seeds of the ecological and human catastrophe now looming in many countries.

The rapidly disappearing face of Kenya.

where I have spent hundreds of nights in camp, and have woken to a sunlit landscape with wildlife spectacles beyond compare. I wish I could think of some reason to help me believe there will still be wildlife left when my children want to take their children to see the finest throngs of large mammals that modern man has known anywhere in the world. From my new home in England, I shall continue to work for the cause of wildlife in Kenya, but I am far from hopeful.

At the same time that I grieve for the wildlife, I feel sadder still for my many Kenyan friends and colleagues. Already they have realized that their yearnings for development are declining with every tick of the clock. With the population growth rate topping four per cent a year, and the economic growth rate slumping below three per cent, Kenya's people are steadily becoming poorer. Many of them now realize it. Some of them recognize that there is little prospect of anything better for years to come. Kenya, like many other countries of black Africa, is no longer part

of the 'Third World' as it is usually defined. Rather these impoverished and deteriorating countries are forming a sort of Fourth World of their own.

In fact, East Africa, the probable site of Man's emergence as a human, is becoming a ravaged region. By the time that modern Africans reach ecological accord with their living space, that is by the time their population growth rate reaches zero (with a population projected to be at least four times greater than today's), the region could be in poor shape indeed. People who read this book are going to witness a brief period during which East Africa will be transformed in a manner that will wring the hearts of all observers. People who want to see East Africa in its pristine glory must go quickly.

This is not to say, to be sure, that the scene is one of gloom and doom from end to end. Southern Sudan, a territory twice the size of France, contains only a few million people – and huge herds of wildlife. Some experienced observers believe that when we count in all

Etosha National Park, Namibia — one of the few countries in Africa where population pressure does not threaten the future of its wildlife.

antelopes of all types, plus zebras and buffaloes and other herbivores, there could be at least two million animals. Among the elephants, there are some of the largest tuskers left anywhere in Africa. Bongo abound (in so far as bongo ever abound), there seem to be fair numbers of cheetah still, and the leopard thrives. Further south, Botswana, twice the size of Great Britain, contains fewer than one million people, with abundant throngs of wildebeest, springbok, gemsbok, and other creatures that tolerate dryish environments. In next-door Namibia, half as large again as Botswana and with only slightly more people, there are some of the most beautiful landscapes I have encountered during my visits to the forty-odd countries of sub-Saharan Africa. Supposing its present political troubles can be sorted out, we can anticipate a reasonably prosperous future there for humans and wildlife alike.

When we turn to Asia, we find, surprisingly and gratifyingly enough, some better news. India's population growth rate is now down to about two per cent a year, and falling steadily – though not nearly so quickly as could be achieved if the Government were to pursue its family-planning policies with sensitivity as well as energy. Pakistan and Bangladesh are not nearly so well placed (in part because of their Moslem culture), with growth rates still around three per cent a year. Bangladesh is a country no larger than England, but already trying to support almost twice as many people; it will have a projected total of 338 million by the time the population stops growing – and unlike England, where 90 per cent of the people occupy only five per cent of the territory, Bangladesh is already bursting at the seams. What hope, then, for the few fine forests that still clothe the flanks of the Himalayas along the country's northern borders, or for the Sunderbans mangrove forests on the Bay of Bengal, largest remaining refuge in Asia for the tiger?

We find a similarly mixed picture in Latin America. Most of the present political upheavals and military conflagrations in Central America are rooted in problems of natural resources. In other words, too many people are trying to subsist off too small a life-support base. The stream of refugees from Haiti and other islands of the Caribbean to the United States has been prompted primarily by deforestation and other forms of gross overuse of human habitats. But these incursions into the United States from the Caribbean are small potatoes as compared with the annual influx of illegal immigrants to the USA from Mexico, at least one million and perhaps two million, fleeing from a land where decades of soil erosion, desert spread, and general overloading of environments are levying their price. The consequences for the United States could eventually prove profound. Experts reckon that by the end of the century, one person in three within the United States could be primarily a Spanish speaker: the very character of US society is being changed by virtue of the declining ecology of its southern neighbour.

Further south, however, there is not quite such a squeeze – not everywhere, not yet. The Colombian highlands, on the one hand, have suffered so much degradation that streams of 'excess' peasants are heading down the eastern Andes into Amazonia – a migration that is actively encouraged by the Government, in order to relieve the pressures of land-hunger. Yet in next-door Venezuela, the Government has virtually forbidden settlement of its sector of Amazonia, and it is able to enforce its edict, partly because people are simply less crowded, partly because the country can depend on its oil wells to foster industrialization and other forms of modernization for its economy, allowing many more people to follow satisfactory lifestyles in urban areas, that is, in a tiny fraction of the national territory.

In short, there is some bad news, and some good news. Overall, the population growth rate of the Third World is now around two per cent a year, and, with a few notable exceptions, declining more favourably than demographers anticipated when they peered into some pretty murky crystal balls in the 1960s. We can be just a little more cheerful now than it seemed we could be in those days when a sizable proportion of bestsellers were about eco-catastrophe.

The Growth of Our Technological Muscle

Homo sapiens, variation *technologicus,* took the stage some 50,000 years ago, when he achieved mastery over fire. Whereupon he rapidly set about the challenge of fashioning the face of the Earth to his own liking. Having been, until that time, almost as much a part of nature as any other organism, he took his first step to set himself apart from nature.

Using the Promethean power of fire, Man began to hunt animals with greater success than he had ever known with crude spears and traps. He could mobilize fire to drive animals over a cliff to their destruction. If he could contrive a fierce enough blaze, he could establish a ring of fire to encircle a herd of, say, elephants. Virtually overnight, he transformed himself from dilettante hunter into maestro hunter.

But it was not primarily through his hunting prowess that his use of fire served to modify the world around him. Many of the fires lit to procure tonight's supper would burn out of control into tomorrow, even into a succession of tomorrows. Woody vegetation in their path would be consumed, and replaced by grass and other herbaceous plants. Hence, according to best scientific opinion, the phenomenon of diminished forests in tropical Africa. In terms of climatic criteria of temperature and rainfall, the forests of the Zaire Basin should extend for several hundred kilometers further into the surrounding savannahs than they do. The answer to the puzzle seems to be that the pyrotechnics of primitive man altered the appearance of a sector of tropical Africa

equivalent in size to Britain and France combined. Not so puny, this unsophisticated creature of human origin.

We can witness a similar phenomenon today, when we go on safari to the Serengeti Plains in northern Tanzania. In 1960, the ecosystem there contained only about 100,000 wildebeest. It also featured extensive woodlands. These wildlife habitats are shared by local Masai stock herders who, to bring on fresh grass for their cattle at the end of the dry season, set fire to the dried-out hay-like cover of the plains. Year by year, these fires burn back the woodlands. As a result there is now much more grassland available for the wildebeest, whose numbers have soared to more than two million, making this one of the most extraordinary assemblies of wildlife anywhere on Earth. Man has certainly changed the ecosystem: has he degraded it, or improved it, or simply refashioned it?

After learning to harness fire for his own purposes, Man moved on to devise agricultural implements which, in conjunction with the crop-growing insights he had developed, allowed him to become a cultivator about 10,000 years ago. This revolutionary advance generated an impact that has left its mark on at least four million square miles of croplands, or one fifteenth of Earth's total land surface – a sizable proportion, when one considers how much is accounted for by deserts, high mountains, ice caps and the like. Moreover, agricultural Man learned to domesticate a number of animals, which in turn required pastureland and other foraging zones, an area that in total grew to the equivalent of twice the amount of cropland. Today we share the planet with at least three billion head of livestock, and, at the rate that cattle, sheep, goats and so on are increasing their numbers, the end of the century may see almost as many domesticated creatures as human beings. Moreover, these multitudes of grazers now tend to overuse their grasslands, with the result that deserts are encroaching at a rate as high as 104,000

square kilometers (40,000 square miles) a year (see Chapter Four).

But it was industrial Man who really took his earthly living space and bent it to his will. He began to mine many minerals in addition to the iron, copper and others that he had long dug from the ground. In particular he started to exploit coal in a big way, followed by other fossil fuels such as petroleum and natural gas. Not that the super-swift exploitation of these minerals is a major factor in itself (although our descendants may one day question whether we put petroleum to its most efficient use by utilizing it to convey us from one supermarket to another). The crucial consequence is that fossil fuels enabled Man to flex his technological muscle to far greater purpose. Using a fuel-driven tractor, a single American farmer can work more land in one week than one of his forebears could cultivate, with his spades and hoes, in a lifetime.

As the Industrial Revolution got under way, so human beings encountered pollution, in the form of smoke from coal burning. Bad as it was, that form of pollution is now being surpassed by a far more insidious and more harmful type of pollution, the spread of toxic compounds. In point of fact, we know little about how large this new threat is, except that hundreds of thousands of kinds of new synthetic molecules are dumped into inland waters and seas each year (see Chapter Six), with the potential impact unknown in almost every case.

Mega-Scale Technology
In recent years, a new and remarkably powerful source of environmental disruption has emerged: mega-scale technology. Within the next few years, we shall probably see a new sea-level canal constructed across the Isthmus of Panama. The strip of land that forms the Isthmus

Fire, man's earliest tool for transforming the environment.

emerged some three to five million years ago, leading on the Atlantic side to a coastal environment with moderate tides, mangrove swamps, sandy beaches and rich coral reefs, and on the Pacific side to strong tides, silt-laden water, rocky shores created by extensive lava flows, limited and impoverished reefs and periodic upwellings of cold nutrient-rich water. The two environments feature roughly 20,000 species of animals and plants, yet perhaps only 10 per cent are found in both regions, less than one per cent in the case of fishes and molluscs. The present canal features a series of freshwater locks that prevent the migration of marine organisms from the Atlantic to the Pacific Ocean and *vice versa*. The new canal would be blasted straight through the mountains, and could allow creatures to wander freely from one ocean to the other. The Pacific communities, having evolved in an inshore environment with marked fluctuations, and having thus developed a higher proportion of opportunistic species capable of wedging their way into existing biotas, might grossly disrupt the Atlantic communities. At the same time, if the Atlantic biotas, with their diversity of species and hence their greater competitive capacity, were to invade the Pacific biotas, as many as 5,000 species

27

Top left: wildebeest on the Serengeti Plain, Tanzania.

Top right: Hereford cattle on the Willicut Ranch, Montana, USA. There are now two head of livestock for every three people in the world.

Above: river pollution in France. Over 60,000 synthetic chemical compounds are in commercial use. A thousand new ones are created each year. Few have been tested to determine their harmful effect on the environment.

could become extinct. If the yellow-bellied sea snake alone were to make its way from the Pacific into the Caribbean, it could inflict so much economic as well as ecological damage that it would be worthwhile to try to prevent any migration at all, through engineering measures such as bubble curtains, ultrasonic screens and intrusions of heated or fresh water.

The Jonglei project to drain part of the Wales-sized Sudd Swamp in southern Sudan is a further example of mega-scale technology. The White Nile flows in a huge semicircle through the swamp, and the project's aim is to bypass the swamp with a 220-mile canal (twice as long as the Suez Canal). Since half the White Nile's flow evaporates while it makes its sluggish way through the swamp, the canal should ensure an appreciable saving of water. Both Sudan and Egypt need lots of extra water. Sudan plans to irrigate 12,300 square kilometers (4,800 square miles) of

land, requiring another ten billion cubic yards of water per year. Egypt, which used to enjoy 23 cubic meters (30 cubic yards) of daily water from the Nile per head of its populace in 1900, now tries to manage with less than a sixth of that, and faces the prospect of only a tenth by the end of the century if its river supplies are not increased. The new canal will drain off 19 million cubic meters (25 million cubic yards) of water per day, or one quarter of the average annual flow, possibly extending to as much again later. The project will reduce evaporation losses by 3.9 billion cubic meters (5.2 billion cubic yards) per year, or one sixth of the White Nile's flow – a considerable saving.

As a result of this diversion of water, part of the swamp will become grassland, which will assist stock-raising. However, a number of estimates suggest that extensive areas of surrounding savannah could become arid, and one tenth of the

The Sudd swamp.

richest pastures seem likely to disappear. No clear indication of the overall environmental consequences is yet available, since the half-billion dollar budget has made only limited provision for pre-project assessment of environmental impact. Nor is any significant investigation envisaged except such undertaken in the course of engineering operations between 1978 and 1985 – almost certainly too late for radical revision of construction plans, should ecological considerations make this advisable. Partly for economic advancement and partly for political security, both Sudan and Egypt seem determined to press ahead with the project, which is financed with Arab oil money and is being executed by a consortium of French and Dutch engineers.

A further purpose of the Jonglei project may be to convert 1.7 million hectares (6,400 square miles) of the zone into a major grain-producing area. As is usual with tropical agriculture, this potential 'bread basket' would attract many new insect pests. The best way to tackle these pests would be to utilize closely associated predators and parasites from the insect communities of the present swamps – and it is highly likely that the Sudd, being a long-established island of moisture amid a semi-arid zone, features an exceptionally rich array of species, many of them endemic. If it turns out that the Sudd's ecosystem is to be fundamentally modified by the Jonglei Canal, whole assemblies of species could become extinct in just a few years.

The same technological clout is being used of course in multiple ways to enhance Man's sojourn on Earth. Through applied science we enjoy a great many more goods and services than would be the case if we had remained innocent of our latent technological genius. In dozens of ways, before we even reach our breakfast table, we find our lifestyles enhanced through the sophisticated items that are available to us at unexpectedly cheap prices. Almost all of us in the developed world live at a level of luxury that would have brought a green glint to the eye of Croesus or Louis XIV. There is no doubt about it – technology has changed our lives from something 'nasty, brutish and short' into something fulfilling and extended.

Fisheries and Forests

As we have seen, however, technology presents us with a Faustian dilemma. We can use it to our benefit, or we can misuse it to our detriment. We have seen how overbreeding of livestock herds leads to overgrazing of grasslands, which give way to deserts. A similarly sad story can be told of the high seas. Until a couple of decades ago, fishermen, even of the most advanced countries, operated as little more than primitive hunters. Now, thanks to sonar-seeking equipment, purse-seine nets and factory ships, fishermen can hunt down whole shoals, indeed entire

Part of the giant bucketwheel being used to dig the Jonglei canal, which will drain the Sudd swamp.

populations, of fish with 'vacuum cleaner' devices that scoop up fish stocks in their tonnes. The result is a gross over-harvest of sea food that should be exploited as an infinitely renewable resource. In the mid 1970s, the global fish catch peaked at a little short of 75 million tonnes a year. Since that time, it has fallen back, recovered a little, and reached a plateau. Many experts believe the fish catch will never go much higher, even though, if we were to use our fishing technology wisely, we could readily take 100 million tonnes year after year. Conservationists in Western Europe and North America are inclined to castigate the Russians and Japanese for killing whales. But the United Kingdom and Norway, the United States and Canada have been just as rapacious in their pursuit of at least twenty species in the northern Atlantic, including cod, haddock, halibut and herring, among other much savoured species. A parallel story can be told of Peru's famous anchovy fishery, which used to register twelve million tonnes of harvest year after year, until it plunged to a mere two million, where it has remained. During the past ten years, a virtually identical pattern of

technological advance, overfishing and steep decline has occurred in both the Gulf of Thailand and the China Sea.

As with fisheries, so with forests. We have already noted the arrival of the giant chipping machine in tropical forests. While temperate forests are somewhat better managed, we find that there is less and less wood to go round throughout the world. In fact the peak, in terms of the wood consumed by each global citizen, occurred as far back as 1964.

The 'Cheap Oil' Illusion

Simply stated, there are now more people, with heartier appetites, and with more zeal to saw off the branch on which they are sitting, than has ever been the case in the past. Before, there has always been some safety valve: new lands to explore, new forests to cut, new mineral discoveries to shore up our material welfare. But now, within a comparatively few years, we have run out of 'slack' in the system – the planetary ecosystem. There is no more 'West' to be won.

In point of fact, the crisis was becoming apparent during the 1960s, but we were given an extension of time

British fishing boat and Russian factory ship — an unholy alliance that has helped deplete North Sea fish stocks.

because of cheap oil. Thanks to petroleum-based fertilizers and pesticides, plus the fuel to run farm machinery and to transport food far and wide, we managed to make ever more productive use of our farmlands. But now that last safety valve is being closed off. When OPEC pushed up the price of petroleum from two dollars a barrel in 1973 to 30 dollars or so in 1980, the oil producers did something more profound than make a trip in the car more expensive. They put an end to the 'cheap cop-out' in agriculture. In 1950, the world's farmers applied fifteen million tonnes of petroleum-based fertilizer to their fields, an amount that increased to well over 100 million tonnes by the early 1970s. That era is ended, never to return.

Other petrochemical products, moreover, have suffered a similar fate. Synthetic fibres, for instance, have been accounting for at least one third of all the clothes in our wardrobes. A not unlikely prospect is that we shall be obliged to purchase more garments made from natural fibres, notably cotton and wool, which, whatever their virtues, will probably prove more costly than the formerly inexpensive polyesters, nylons and other petroleum-derived textiles.

The Hamburger Connection
A theme being developed throughout this chapter is the international linkages between lifestyles in one part of the world and environmental repercussions in other parts of the world. These linkages, which are likely to become ever more pervasive and important in the future, can be illustrated through the 'hamburger connection'.

As we have noted, tropical forests are being steadily depleted in all three major regions of the biome. In Central America, the main agent of destruction is the cattle raiser, who clears away a patch of forest in order to establish artificial grasslands. In rough terms, he can be reckoned to be eliminating some two million hectares (8,000 square miles) of forest per year – and his impact is increasing more rapidly

The hamburger connection – South and Central American cattle, via McDonald's, the world's largest seller of hamburgers, to millions of mouths. In the process millions of acres of tropical forest are being destroyed.

31

than that of the commercial logger.

The cattle raiser's activities are largely stimulated by consumer lifestyles in affluent sectors of the global community. As beef produced in developed nations (and especially in North America) becomes more expensive, the rich-world consumer fosters the spread of cattle raising into the forest zones of Central America. He does not do it wittingly, and certainly not with wanton intent, but he does it effectively and increasingly.

During the last two decades, consumers in the developed nations have revealed an apparently insatiable appetite for beef. In 1960, an average American consumed 38 kilograms (85 pounds) of beef per year; by 1976 that total had risen to 59 kilograms (135 pounds). According to projections prepared by the Food and Agriculture Organization of the United Nations (FAO), the demand for beef will continue to rise more rapidly than for any other food category except fish, until at least 1990.

At the same time, few items in the shopping basket have increased more in price during the past few years. In the United States, beef prices have been soaring far faster than the overall cost of living. Between 1975 and 1979, the price paid to farmers for a Montana steer rose from $0.29 per pound to $0.48; in the same period, retail prices of beef soared by 30 per cent per year, until by 1980 they had topped $2.25 per pound. (One pound is equivalent to 0.4536 kg.)

Not surprisingly then, Americans have been seeking low-cost beef elsewhere, and Central America has proved to be a ready supplier. In 1978, the average wholesale price of beef imported from Central America was $0.67 per pound, compared with a wholesale price of $1.50 for similar beef produced in the United States.

Cattle in Central America are raised on grass rather than grain; so the beef is very lean. This makes it suitable for only one sector of the US beef market, the fast-food trade. According to the Meat Importers Council of America, virtually all the Central American beef makes its way into hamburgers, frankfurters, luncheon meat, hot dogs and other processed meat products. As it happens, the fast-food trade is the fastest growing part of the entire food industry in the United States. Throughout all of the 1970s, it grew at a rate of 20 per cent per year, or two and a half times faster than the restaurant industry overall, and Americans now spend five billion dollars annually for fast food. One fifth of the US food budget is spent dining out, and that proportion could rise to one half by 1990; the demand for cheap lean beef is unlikely to slacken.

Faced with one inflationary price rise after another during the past several years, the US Government has repeatedly stepped up beef imports. In 1960 the USA purchased virtually no beef abroad; it now imports 800,000 tonnes each year, or around 10 per cent of its total consumption. Of these imports, 17 per cent comes from tropical Latin America and three quarters of this (just over 100,000 tonnes) from Central America. The additional imports in recent years have been mainly from Central America, as well as from other sources such as Australia. And although the extra purchases from Central America contribute less than one per cent of the US consumption of beef, the Government estimates that they supply enough additional cheap meat to trim five cents off the price of a hamburger each year. Food is the sector of the US economy most susceptible to inflation, and the greatest price increases have been for meat, and especially beef. It is believed that increased imports of cheap Central American beef have done more to stem inflation in the US than any other single Government initiative. Indeed, the US Government calculates that the additional imports save consumers at least 500 million dollars a year.

Since 1960, the area of man-established pasturelands and the number of beef cattle in Central America have both increased by two thirds. The pasturelands have almost all been established at the cost of previously undisturbed rain forests, which have been reduced by almost 40 per cent since 1960. At present rates, all remaining forests will have been eliminated by 1990 or shortly thereafter.

In terms of their biological diversity, these rain forests constitute some of the richest ecosystems on Earth. For example, Costa Rica, a country hardly bigger than Denmark, has 758 bird species, 620 of them residents, or more than are found in all of North America north of the Tropic of Cancer. Costa Rica likewise harbours over 8,000 plant species, with more than 1,000 orchids. The La Selva Forest Reserve, amounting to only 1,800 acres, contains 320 tree species, 42 fish, 394 birds, 104 mammals (of which 62 are bats), 76 reptiles, 46 amphibians and 143 butterflies – a tally that is, broadly speaking, half as many again as California's.

It is ironic that if Central America's forests disappear within the foreseeable future, not only local people will suffer by way of environmental degradation, decline of watershed services and the like. (Already Costa Rica's hydropower dams are being silted up, and Honduras has undergone increased hurricane damage because of the loss of forest cover.) Other human communities will suffer, notably the North Americans. Central America's rain forests, with their exceptionally rich biotic diversity, contain many genetic resources of great value to modern agriculture, medicine and industry. For example, in 1978 a wild variety of perennial corn was discovered in a forest of southern Mexico. Not only could this new strain enable the corn-growing industry, through cross-breeding, to avoid the season-by-season cost of ploughing and sowing, but the wild germplasm offers resistance to several viruses which attack commercial corn. And according to a botanist from South Carolina, Dr Monie S.

Hudson, who specializes in medicinal applications of phytochemicals, a screening project to evaluate 1,500 tree species in Costa Rica's forests has revealed that around 15 per cent might have potential as a treatment for cancer.

It is clear then that both Central Americans and North Americans are contributing to the 'hamburgerization' of the rain forests, and that both will suffer from their loss. It is equally clear that both must co-operate if the problem is to be solved: either all will lose together, or all could gain together. This is a paradigm of interdependent relationships of resources within the international community.

Additional Economic–Ecological Linkages
In a similar style, the demand patterns of the rich world and the supply patterns of the developing world combine to cause the degradation of savannah lands in Africa. Since there is still unranched savannah left in Africa (just as there are 'unhamburgerized' forests left in Central America), livestock owners in Africa tend to build up their herds of commercial cattle by extensive (rather than intensive) modes of livestock husbandry. That is, they expand into savannah grasslands that have hitherto remained little used. At the same time, the stockman finds he has to operate on increasingly narrow profit margins, due in part to international market constraints. These twin factors – increase in herd sizes, and increase in stock-raising pressures – leave stockmen less ready to share savannah grasslands with such wild creatures as zebra, antelopes and the like, together with their predators such as lion, leopard and cheetah. The result is a steady attrition of wildlife numbers. The expansion of commercial stock-raising in Botswana, for instance, stimulated as it partly is by marketplace demand in Western Europe, is a main cause for the cheetah's decline in Botswana. Botswana now dispatches 80 per cent of its beef output overseas, including 30,000 tonnes a year to Western Europe. When a European shopper goes

The tropical rain forest of Costa Rica, one of the world's richest ecosystems.

to the local supermarket to buy beef that comes from Botswana, she or he wants it at a 'reasonable' price, that is, at a price competitive with beef from other sources that may not feature wild predators as a major cause of stock losses. She or he thereby encourages – unwittingly and without malice toward African lions or leopards or cheetahs – the cattle raiser in Botswana to eliminate wild predators on his savannah-land holding.

Similar pressures induced by the lifestyles of the advanced world are causing African cultivators to migrate into the savannah grasslands. The flood of landless people heading for wildlife habitats could be partly stemmed by more favourable trade terms in the international markets. For example, a Tanzanian farmer must now sell five times as much coffee to buy a foreign-made piece of machinery, such as a light tractor, than he did in 1960. He can earn a tolerable cash income for himself and his family from two or three acres of coffee in the more productive parts of Tanzania, but he must cultivate between three and five times as large an area in savannah zones, which he must plant with grain crops rather than coffee, to ensure the same income. In so far as he usually receives little more in real terms for his coffee than in 1960, he sees diminishing cause to stick with the cash crop. As a consequence, he is increasingly inclined to abandon attempts to make a living off a small patch of high-productivity land, and is likely to join the throngs headed for the savannah grasslands.

Furthermore, the greatest share of the profit from the most widespread form of coffee (powdered coffee) accrues after the raw beans have left Africa. If processing of coffee beans were permitted before export from Africa, the crop could be made much more profitable for Africa's coffee growers. But the multinational corporations that dominate the coffee trade generally refuse to purchase coffee in any form other than raw beans, so that they can retain the high-profit processing. In this sense, the life-support systems of

African wildlife extend to the breakfast tables of Europeans and Americans, where coffee is the commodity which, apart from the temporary boost in prices of the mid-1970s, has shown the least price increase during the past 20 years.

Much the same analysis applies to consumers in Western Europe and North America when they purchase other goods from developing Africa, such as tea, cotton cloth and textiles, sisal and the like, all commodities which are subject to marketing systems similar to that of coffee.

The fate of African environments is thus determined not only by local circumstances. It is influenced, in part at least, by the lifestyles of the developed world. These economic–ecological linkages between different members of the international community are little recognized to date, but they represent a significant factor for land-use patterns in Africa's savannahs.

Scope for Action

A time of crisis, as the Chinese say, is a time of opportunity. It is a time when the old order changes, when the established moulds are cracked, when conventional ways of doing things are discarded. Plainly we are going to have to respond, sooner or later, to the problems of an overloaded planet: either sooner, through safeguard measures of sufficient scope, or later, when we find that the degradation of our ecosystems represents a loss through which the human community is universally impoverished. In a world that is running short of one resource after another after another, there is one resource that we have scarcely tapped: the human ingenuity to do things better. So let us not talk so much about problems. Let us regard them as challenges. Will we prove equal to an era that demands we walk ten feet tall?

Norman Myers

Chapter 3

The Changing Face of Agriculture

We live in a world where untold millions of human beings go hungry. It is not that they miss a meal now and then. They are not dieting or fasting. They are hungry, every day of their short lives, because they cannot obtain the food they need to keep body and soul together. Our emotions are assaulted by the pictures we see of deformed, crying infants, doomed, we are told, to succumb quickly to illnesses their hunger has made them too weak to resist. They cling to their mothers, who try to retain some human dignity amid the deepest despair of which we can conceive. The tragedy of the world is written on the faces of the poor of the world. It is a tragedy of appalling dimensions.

Such images, brought insistently before us by the global network of mass media, can lead to a sense of despair. With so much suffering, so many mouths to feed, where can we make a start on bringing relief? If the hungry, like the poor, are always to be with us, the sheer scale of the task may intimidate us, sapping the will to work towards a more decent world.

It need not be so, however. It *is* possible to think about the problem in ways to make the task more manageable.

Catalogues of Want

First, we must distinguish between two types of mass hunger: that caused by a sudden, calamitous disruption of the food supply, and the longer-term famine caused by repeated crop failure in regions too poor to cope.

The essential difference is that short-term hunger can occur in any society, rich or poor – although it is true that it does so more often in the poor countries. When disaster strikes, voluntary agencies are well-equipped to get food, blankets, tents and medical aid to the stricken area. Ordinary people in rich societies respond quickly and

Top: Calabrito, Italy, 1980. Natural disasters, such as this earthquake, can strike anywhere. Here relief was quick and reasonably effective — in most of the world it is neither.

magnificently with donations to support the relief agencies. When relief goes into a society with a solidly established system of services – hospitals, doctors, supplies, transportation – the emergency is usually over rather quickly. In poor countries without such services relief may take much

longer but, as long as the money holds out, short-term calamity can be dealt with.

Long-term famine, however, occurs only in poor countries. It was not always so: little more than a century ago the potato famine halved the population of Ireland. As we shall see, high-technology agriculture in

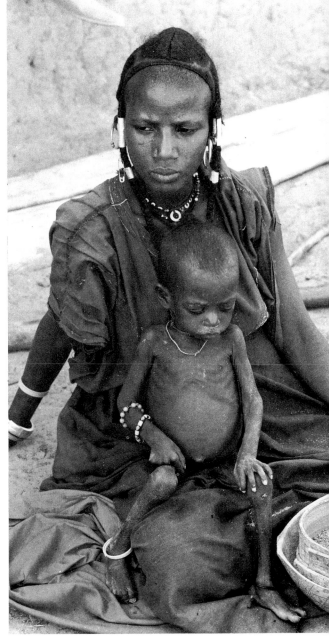

Above: Cambodian refugees in Thailand. The ancient response to famine is to move. In our more crowded world even this desperate option is less available.

A quarter of the world's population, over 1 billion people, share the plight of this woman from Upper Volta and her child. They are always hungry. Half of them are starving.

rich countries has come to protect them from famine by exporting it to the Third World. For the moment, however, it is important to understand that repeated crop failures in poor countries begin a vicious circle of problems which only poor countries have. Allied with the lack of services already mentioned is the fact that the land itself is poor and liable to frequent drought. Once, populations inhabiting this marginal type of cultivation could simply move on when their crops failed. Today, political divisions mean that starving populations have literally nowhere left to go, and no alternative except to die of hunger on the barren land which will no longer feed them.

The Anatomy of Famine

When famine strikes, it stays. The crops fail in one year. That does not mean the farms produce nothing at all, but only that yields are well below those of non-famine years. From every crop a proportion of seed must be kept in reserve for planting the following season. If hunger is severe, however, people may eat their seed stocks just to stay alive. When this happens there is less to sow the following year, so yields remain low, and the famine perpetuates itself. It can take several years to recover from the effect of a single bad year.

When famine strikes, not everyone goes hungry. There is some food, and so some people eat. The price of food is determined by demand and supply. If the supply is short but the demand remains constant, the price rises. Only the wealthier people can then afford to buy it. Famine kills only poor people.

Exporting Famine

Today, rich countries have eliminated famine. They have found 'food security', for which the source is mainly the United States. The United States is not the largest producer of food in the world, but it is the largest exporter of food. It has such a large area of cropland that without much effort it can produce far more food than its own people can eat. Countries that are short of

Top: a US grain ship loading in Philadelphia. The USA exports nearly 100 million tonnes of grain each year. The next largest exporter sells only a fifth of this amount.

food import it, and since the 1950s the amount of food being traded around the world has increased dramatically. The actual commodities that are traded are principally cereal grains, of which the total volume and value are so large as to dwarf other food commodities into insignificance. Grain markets that operate internationally may trade in food that has not yet even been grown, on estimates of the size of the next crop.

If food is in short supply internationally, its price rises. The rich countries continue to buy it, paying more, but poor countries may be unable to afford it. The extent of the world trade in food is such that the entire globe now resembles a single country of former times, in which the rich countries behave as rich people used to behave, and poor countries find themselves trapped as poor people used to find themselves trapped. The consequence is depressingly simple. A crop failure anywhere in the world may lead to famine, and that famine occurs in the poor countries. We have learned to export famine.

Even so, food is seldom cheap. The farmers who grow it must be paid, or they will go out of business. If food is so plentiful that the price falls below the cost of production, governments must intervene to subsidize their farmers, and the taxpayers must pick up the bill – provided, that is, the taxpayers are rich enough. When food is plentiful, therefore, and the granaries are full, the United States Government pays some of its farmers to let their land lie fallow. The farmers are paid to grow less. This reduces the amount of food in the world, maintains prices, and keeps agriculture in production. At present, in 1983, world surpluses of food are large and American land is being taken out of production. The stock of food held in reserve amounts to about 250 million tonnes, and despite American moves it is likely to grow further. Canada is expected to export, or to try to export, 27 million tonnes of grains, Australia 9 million, Argentina 19 million, the EEC 21 million, and the USA 96 million, making a total of 172 million tonnes. Against this, the

Above: food aid, such as this sorghum being bagged for distribution in Dakar, is a bitter-sweet gift. Whilst providing much needed immediate relief, it undercuts the efforts of local farmers to provide a long-run answer to food shortages.

Soviet Union is expected to import 33 million tonnes, Japan 24 million, China 16 million, and the EEC (which imports certain types of grains and exports others) 12 million, making a total of 85 million tonnes. The difficulty is obvious. Farmers are growing food that no one will buy. The poor will not buy it. They cannot afford it, for the producers must be paid enough to allow them to live as others live in rich countries, and they must be paid in currency they can use, which to the poor is foreign currency.

Food Aid

How can there be huge surpluses of food and many hungry people all at the same time? Surely it should not be beyond anyone's wit to move food from where it is not needed to where it is? True, the farmers have to be paid, but if the poor cannot afford the price then perhaps the rich will pay it for them – what is the use of being rich if you cannot help those in need?

This superficially generous idea was tried. The experiment began in 1954, with an American project called 'Food for Peace', which was copied by Canada, France and Australia. Food was supplied to needy countries free, or for a very low price, or just for the cost of shipping it. These schemes ended finally in 1970, by which time the USA had moved nearly 180 million tonnes of grain.

It was not entirely altruistic, for there were conditions. The most important was that a recipient country should not increase its own exports of commodities that competed with US products. Burma, for example, used to export rice, but it was not permitted to use the food aid it received to increase its rice exports. It could not eat the free food and sell its own produce. That may sound a very reasonable condition, but life is never simple. Burma was poor, it needed foreign exchange, and rice was the only commodity it could use to obtain it.

The programmes failed, in fact. They supplied food much more cheaply than food could be produced by farmers in the recipient countries. Those farmers, too, had to be paid. They had to be able to live, to

support their families, to buy the non-agricultural products they needed, and to buy the materials essential to continue farming. Always, everywhere, there is a minimum price farmers must receive for their produce, below which the amount of farm produce actually diminishes. The food aid was meant to be given to the very poor, but that is not what happened. It entered the ordinary markets, undercut local prices, and farmers suffered. If one rich country tried to distribute very cheap imports in another rich country it would be called 'dumping' and it would be stopped very quickly. Dumping is dumping wherever you do it, and by 1970 it was clear that had they continued, the food aid schemes would have so depressed the agriculture of the

Rosa Grande, Nicaragua, 1980. These Sumu Indians are practising one of the oldest forms of agriculture – slash and burn. A clearing cut in the forest is burned to release nutrients into the soil. The crops are then planted.

recipient countries that those countries would have become dependent on such aid in perpetuity.

The 'Green Revolution'

Clearly there was a need for some alternative approach. Perhaps more land could be brought into cultivation and food output increased that way? Certainly this seemed possible. In the world as a whole much less than half the land area that could have been farmed was being farmed in the mid-1960s, and much of the 'spare' land was in poor countries. In Africa, for example, less than one quarter of the potential farmland was being used, and in South America little more than 10 per cent was cultivated. In Asia, however, there was much less room for expansion.

Agriculture has extended into new land, especially into marginal land where the structure of the soil and the climate are ill-suited to it. There are three types of land where farm expansion is likely to cause problems: steep hillsides, land in the humid tropics that supports forest, and land close to the edge of a desert.

On steep slopes the removal of an established vegetation cover often leads to severe erosion of the topsoil, as rains wash away particles that formerly were bound together by roots. This is especially the case in mountainous areas where the melting of winter snows at higher levels causes a large increase in the amount of water that flows across the surface, and where rainfall is

41

markedly seasonal.

In the humid tropics, on more or less level ground, the effects of clearing the established vegetation are different. Although tropical rain forests are luxuriant, they grown on very old, poor soils. Decomposition of waste matter is so rapid that the nutrients on which life in the forest depends are largely contained in the living plants and animals themselves. When these are removed there is little reserve in the soil to sustain cropping, and attempts at conventional farming usually fail after a few seasons. Where plant growth is possible it is often so prolific that weeds and pests defeat the farmers.

Close to the edge of a desert, interference with fragile soils is likely to exacerbate the spread of the desert (see Chapter Four).

These are aspects of agricultural expansion that are described in more detail by other contributors to this book and there is no need to dwell on them here. There remains plenty of uncultivated land into which farming might spread without causing damage to these especially sensitive areas, certainly if the ecological consequences of such expansion were taken into account by the planners. They may also need to consider the consequences of intensifying certain existing methods without allowing them to be used in new areas.

The 'slash-and-burn' type of primitive farming has been practised for centuries. Today we think of it mainly in association with the tropics, but soon after the end of the last glaciation it was practised over most of Europe, and it cleared away much of the original forest. Usually the farmers choose a hillside where they then fell selected trees in such a way as to make them bring down others as they fall. Having removed such timber as they need, they wait for foliage to wilt, then fire the area. After it has cooled they plant their crops in the spaces they have cleared. The soil is temporarily fertile, but in time its fertility drops and the forest begins to regenerate. Eventually the area is abandoned and a new area is cleared. A group of such farmers works in a kind of

AGRICULTURAL PRODUCTION POTENTIAL

This map illustrates the maximum production which might be possible if the agricultural potential of all our land were fully used. Cost and competing uses for land will substantially reduce this potential.

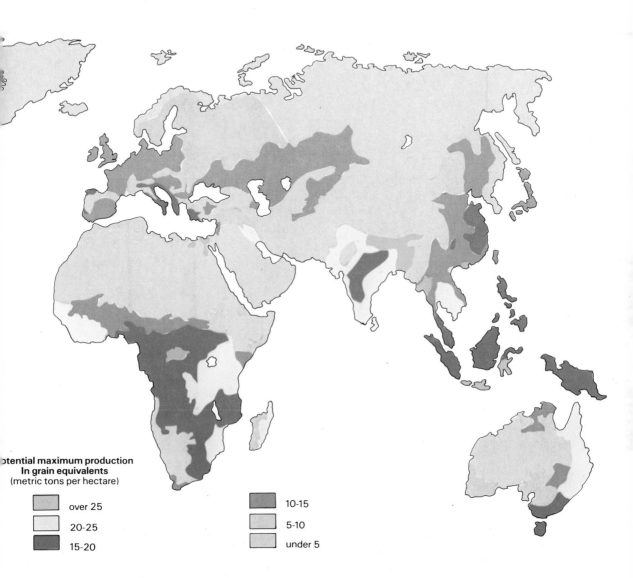

Potential maximum production
In grain equivalents
(metric tons per hectare)

- over 25
- 20-25
- 15-20
- 10-15
- 5-10
- under 5

rotation, returning to each site every eight to twelve years depending on the region, and the system can be continued for long periods. It does use up a large area of land, however – especially if the sites are included that are not being tilled but to which the farmers will return – and so it is a very unproductive system. When the total land area is reduced, because people are practising settled farming on some of it, or because there are just too many primitive farmers, the cycle accelerates. Sites are revisited more frequently, soil fertility is not restored fully after each visit, so the cycle accelerates further, and in time the area is reduced virtually to desert. That drives the farmers deeper into previously untouched land. Slash-and-burn farming is one of the major causes of tropical forest clearance today.

Expansion into new lands was rejected as a strategy by students of the problem for purely economic reasons. The fact was that farmers were not producing enough from the land they already farmed. If they expanded into new land the resources for investment in agriculture would be spread more thinly and the result might well have been an overall drop in output as a larger area was farmed to even lower standards.

The FAO, based in Rome, analysed the problem and decided that the only practicable way to relieve hunger was to increase food output on existing farms; they devised a complex programme of technologies and strategies that might achieve this. The plan was published in 1962 as 'The Indicative World Plan for Agricultural Development'. This was too big a mouthful for the popular press, too dull and bureaucratic-sounding for a document dealing with so dramatic a problem. Someone, no one knows who but almost certainly a journalist, nicknamed the plan the 'green revolution', and the name stuck. We must live with that name, because it exists and it is catchy, but it is very misleading. Although some officials may have been deceived by their own enthusiasm, it never was likely that deserts would bloom, or that hunger would

Intensive agriculture, with its use of chemicals, machinery and new crop varieties, has vastly increased the yields from croplands. But at a price. In the 40 years since World War II the appearance of large areas

*of the English countryside has been totally altered.
Gone is the familiar patchwork of small fields
separated by hedgerows (top) and in its place are large
open fields reminiscent of the American*

*prairies (right). Over 120,000 miles of hedgerows
have been uprooted in the process (left) destroying one
of Britain's few remaining habitats for wildlife.*

disappear overnight. There was no plan to make 'green' areas that were not green already, and although it was far-reaching the plan promised no revolutionary improvement. It called for years and years of hard work, investment, dedication, and, with luck, a slow but steady increase in the amount of food available to the poor.

The plan was intricate. It ran to more than 670 pages, and had a separate 72-page volume just to summarize its recommendations. This was no brief summons to the barricades and it contained no snappy slogans, only chapter after dreary chapter of close argument, numbers and jargon.

Miracles

The key ingredients of the 'green revolution' plan were the newly developed 'high-yielding varieties' of wheat and rice. Again, the name was inappropriate, for these varieties were high-yielding only under certain specified conditions.

Essentially, the aim was to produce in low latitudes the change that had transfigured farming in the higher latitudes mainly during the 1950s and 1960s. It was achieved mainly by the investment of capital and by guaranteeing the prices paid to farmers; with this security came a rapid technological advance. Farms were mechanized, allowing the land to be worked more quickly while conditions were suitable, and reducing crop wastage. Chemical fertilizers were used much more to boost yields. Pesticides were used to reduce infestations by disease, pests and weeds. At the same time the workforce decreased. Between 1955 and 1966 in Britain the farm labour force fell by 35 per cent. During the same period crop yields increased by about 50 per cent. Farms were amalgamated to conserve capital, and the face of the countryside was altered.

The plan took full account of the special difficulties of transferring improved farming methods to non-industrialized economies. The problems were: maintaining agricultural employment; adapting the use of fertilizers to conditions of low, or high, rainfall, and developing new

WORLD GRAIN PRODUCTION 1950-80			
Year	Population	Grain Production	Grain Producti Per Capi
	(billions)	(million metric tons)	(kilogram
1950	2.51	631	251
1960	3.03	863	285
1970	3.68	1137	309
1971	3.75	1237	330
1972	3.82	1197	314
1973	3.88	1290	332
1974	3.96	1256	317
1975	4.03	1275	316
1976	4.11	1384	337
1977	4.18	1378	330
1978	4.26	1494	351
1979	4.34	1437	331
1980	4.42	1432	324

Source: U.S. Department of Agriculture and United Nations

varieties of crops best suited to local conditions.

For example, suppose that water and fertilizer are available. Apply them to a cereal plant and the plant will grow taller, and its ear will grow heavier. The extra weight of the ear often exerts such leverage on the weakest part of the longer stem that the stem breaks. The ear falls to the ground, where it may be impossible to harvest it and where in any case it may be wettened and the grain in it may germinate. This process is called 'lodging', and in temperate countries new varieties of cereals were bred that would not lodge, because instead of responding to fertilizer by growing taller, they responded by producing more stems,

At the heart of the 'Green Revolution' are new varieties of cereals adapted to give high yields in the climatic and agricultural conditions of developing countries. As a result, food production has risen rapidly, particularly in Latin America and East Asia.

and therefore more ears, on a dwarf plant.

Unfortunately, these varieties could not tolerate the warmer growing conditions in low latitudes. Work on this problem began back in 1943, in Mexico, where Dr Norman E. Borlaug was leading a research project sponsored by the Mexican Government and the Rockefeller Foundation. By 1954 some progress had been made, but not much. Yields could be increased, but lodging remained the major limiting factor. Then they imported a short-strawed Japanese variety of wheat, crossed it with varieties on which they had been working, and within a few years they produced their new, short-strawed variety. It produced a crop of 15 to 20 tonnes a

hectare, which was more than double the best yields they had obtained previously.

This was the first of the 'high-yielding varieties' of cereals and soon it was followed by rice. Bred at the International Rice Research Institute, in the Philippines, 'IR8' was the prosaic name given to the first of the series of new rice varieties to go into commercial use. It was ready for use in 1966, could yield up to 6.5 tonnes per hectare (which is about three times more than traditional varieties), and it matured much more quickly, so that two or even three crops a year could be obtained from land that formerly produced only one. It was a small plant, growing only about one meter tall. This was the rice the press nicknamed 'miracle rice'. It is not difficult to see why.

The 'Green Revolution' Turns Red

At a casual glance, then, it looked in the mid-1960s as though the problem of world hunger had been beaten. Indeed, some people began to warn of a world glut of food that would replace the world shortage.

A casual glance was as much as most people gave it. Everyone wanted a solution to be found and it was tempting to gloss over the less reassuring details. For example, farmers could not obtain the seed and fertilizer they needed. If they could obtain it, in many cases they could not afford to pay for it. Banks were unwilling or unable to lend to poor peasants. Sometimes farmers would grow the new varieties only to find people did not like to eat it. The rice in particular did not cook well.

The plan required that peasant farmers be brought for the first time into a cash economy. They had to buy and sell. They had to buy seed, because no longer could they retain seed from one crop to sow for the next. The new varieties were hybrids, exploiting the vigorous growth that occurs in many hybridized plants, but they were unstable. If sown from their own seed they tended to 'unscramble' themselves into the varieties from which they had been bred. Indeed, the entire package was necessary, for without adequate water and fertilizer the new varieties yielded little more than the old ones.

Finally, difficulties arose because of the power of landowners in agrarian societies. We have forgotten just how powerful the rich farmer, the squire, the feudal lord, used to be in the developed countries, and how difficult it might have been for a peasant to introduce major reforms. Stories were told of villages in which only the rich farmers could afford the new technologies. They produced higher yields, which reduced prices, which forced the poor farmers from the land and left their holdings in the possession of the big landowners. There were farmers who refused to increase wages when labourers had to work harder to bring in larger harvests. There were tenant farmers whose rents were increased to deprive them of the increased profit from the new varieties. In some cases landlords simply took back tenanted land, reducing their former tenants to the status of hired labourers and even driving them from their homes, which occupied valuable space. The journalists, never at a loss for a catchy phrase, wrote of 'the green revolution turning red', and the politicization of rural communities was indeed one product of the changes.

Critics grew more vocal. Where a situation is as grave in immediate, human, understandable terms as that relating to the need and supply of food around the world, it is easy to slide from over-optimism to over-pessimism. People had seized on slogans that allowed them to forget the problem, to assume it had been solved. When years passed and very clearly it had not been solved, they assumed that the programmes had failed.

The truth was that although the modernization of farming in the poor countries fell short of its targets, and continues to do so – not least because it failed to predict the rise in oil prices and their effect on fertilizer prices – it has achieved a great deal. On average, agricultural production has been increasing by nearly three per cent a year for the past twenty years. It is not enough yet, but it is sufficient to have doubled output inside 25 years, and that rate of increase is comparable with what was achieved in Britain up to about 1960, since when the rate of increase has slowed. By any reasonable standard, the 'green revolution' must be counted a success.

The trouble is that hunger continues. Success was not enough.

The Physiology of Hunger

Perhaps the campaign would have proceeded a little faster, and would have achieved a little more, were it not for a fundamental error that coloured it for some years. The error was summed up in another slogan, 'the protein gap'.

The hungry people who were seen and photographed, and especially the hungry children, often showed signs of protein malnutrition. When their diets were analysed they were found to contain much less protein than the diets of people in the rich countries. It was concluded, therefore, that their hunger was largely a matter of protein shortage. Strenuous efforts were made to find ways of increasing the amount of protein in their diets. This is possible, but it is difficult and expensive. It is expensive because protein-rich foods cost more to produce than carbohydrate-rich foods, the foods rich in sugars and starches. Meat, eggs, and cheese are typical high-protein foods, and they come from animals, which eat vegetables. An animal must eat about ten kilograms of plant protein to produce one kilogram of animal protein. Humans cannot rely on the plant proteins directly, it was argued, because they are deficient in some of the ingredients that we need although herbivorous animals do not.

The first efforts made to close the protein gap were based on the industrialized processing of plants to yield high-protein foods. They failed because the foods were either unfamiliar to, or unsuitable for, the diets of those who were meant to benefit.

The efforts continued nevertheless, but it was found that when the protein-starved people were fed protein-rich food, they went right on being protein-starved. It was then that the fallacy was revealed.

The human body needs protein foods and energy foods. The protein is needed for growth and the repair of tissue, but it cannot be so used, indeed it cannot even be digested, unless energy is available. In fact without energy we die very quickly, and all the protein in the world is useless, at least as protein. The body will obtain its energy in any way it can and if there is no other source it will break down protein, rather inefficiently, and 'burn' it. If you eat too little for long enough, your body will begin to use the protein in your food just to keep itself going, and you may well start to show signs of protein-malnutrition. If you eat more protein, that too will be used to provide energy, and it will go on providing energy until you are eating so much that there is food to spare and protein can be used as protein once more.

The Ecology of the 'Green Revolution'
The blitz on protein deficiency proved a distraction from the main issue. Meanwhile, it was being discovered that the 'green revolution' had hidden costs to weigh

Buying seed in North Nyanga. One effect of the Green Revolution has been to bring many peasant farmers into the cash economy for the first time. Protein is essential for a healthy diet, but carbohydrates (energy foods) are even more so. It takes 10 kg of grain to produce 1 kg of meat. Grain-fed cattle are thus a very wasteful way of supplying protein.

against its promised benefits. When apparently fertile virgin land was brought under the plough, it might become more vulnerable to climate. This happened in the low-rainfall regions of Oklahoma and the 'virgin lands' of the Russian steppes. When the surface vegetation was ploughed up, no roots remained to bind the soil. It dried in the hot summers, and the prairie or steppe winds carried it away, creating dustbowl conditions. In such cases the fertility of the soil declines and the land may eventually have to be abandoned.

In the tropics, heavy rainfall may wash away newly cultivated soil, carrying with it seed, fertilizers and the nutrients which made the soil fertile in the first place.

A more subtle danger of high-technology farming – vulnerability to attack by pests or disease – is associated with monoculture, that is, the widespread cultivation of a particular crop variety. Areas with a large number of farms growing many different varieties of a single crop have an inbuilt immunity to large-scale destruction by pests or disease. For example, there are microscopic fungi which attack wheat. Such an organism is likely to be highly selective, preferring one variety above all others. So, although some fields will be wiped out, others, growing different varieties, will survive.

Let us suppose now that a scientist wishes to confer immunity to that disease upon the variety of wheat it has attacked. He will do this by examining that variety, and the other varieties, to discover why the disease is so selective. Then he will cross breed the varieties, to try to introduce the characteristics that confer immunity from immune varieties into the susceptible variety.

You will see that a multiplicity of crop varieties helps everyone. It helps the farmers, their families, and it also helps the scientist who arrives to help them.

Now consider what happens when a new variety is introduced that yields two or more times more heavily than the old varieties. No farmer can afford not to grow it. So before long the new variety takes over

THE DANGERS OF MONOCULTURES

Top: Oklahoma in the 1930s – the 'Dust Bowl'. Immortalized in John Steinbeck's 'The Grapes of Wrath', the dust-bowl years were both a human and an ecological disaster. Wind-blown soil from the farms of Oklahoma fell on to ships 300 miles out in the Atlantic.

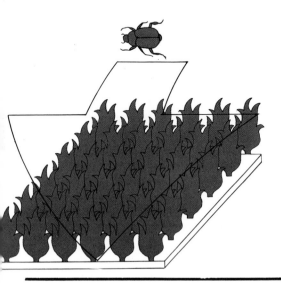

and is grown everywhere. All the old varieties are abandoned.

What happens now when the disease discovers that the new variety is very much to its taste? All the fields are filled with it. The old varieties no longer exist to provide a buffer against failure, or breeding stock for the scientist. The entire region becomes more prosperous, but at the same time extremely vulnerable to ill fortune.

Pollution Problems

Using fertilizers can pollute the water. This might have been a serious problem in the 'green revolution' areas. In fact it has not been, but this is due more to the shortage of fertilizer and the relatively meagre amounts used than to good planning.

Pesticides, too, may cause pollution, but again the problem has been less severe than might have been anticipated. The reason is much the same, but only up to a point. Because both the chemicals and the equipment needed to spray them, not to mention the fuel used by the sprayers, are expensive, a cheaper way of using them had to be found. The solution lay in 'ultra-low-volume' sprayers. Made from plastic, powered by torch batteries in the handle, carried by hand, such a sprayer delivers its chemical not as a drench from nozzles but as a fine mist, composed of droplets all the same size. The mist drifts almost horizontally into the crop, borne on the wind. It is not only cheap, but much more efficient than the more conventional sprayers used on farms in the rich countries, and it uses one tenth to one hundredth the amount of pesticide. Pollution problems are fewer, because the amount of pesticide being released into the environment is less.

The supply of water has caused problems, however: very real ones. Irrigation must be accompanied by drainage, but digging drains adds greatly to the cost. It is much cheaper simply to divert water from a river and 'water' the land. What is more, it will appear to work very well for quite a long time. As is explained elsewhere in this book, however, such primitively irrigated areas can later on become

Above: many plant diseases and pests have very specific requirements for their food. Thus they will attack only one variety of a plant such as wheat. Traditional agriculture mixes varieties, thus each field will have some resistant and some vulnerable plants (left). If the crop is attacked, only some of it will be damaged. Intensive agriculture plants whole fields, sometimes whole districts, with exactly the same variety. If the crop is attacked, then the whole of it may be lost.

waterlogged or salty, so that plants cannot grow. It is estimated now that the area brought under irrigation each year is roughly equal to the area lost through waterlogging and salinization caused by inefficient former irrigation.

The remedy is expensive. Crops may be grown that tolerate the conditions until, in time, the water-table falls and the rains wash the salts from the soil. More usually, drainage must be provided to remove the surplus water, more irrigation water must be poured on to the land, and the salts must be leached away using valuable water that is really needed elsewhere.

Blame the Customers?

The original attraction of the 'green revolution' was the simplicity of the whole enterprise. Rational application of Western skills and resources would solve the problem of world hunger. This, at least, was the popular view. When the solution began to bristle with uncomfortable difficulties, popular opinion – led by journalists and politicians – turned to other 'rational solutions'. Birth control seemed an obvious candidate, with the added attraction that it put the responsibility for hunger squarely on the shoulders of the hungry.

Naturally we can support this line of argument. We can show that impressive though it is, the rate of increase of agricultural output barely matches the rate of population growth (*see* Chapter Two). What is more, we can explain the population

growth. Improved medical care has increased the survival chances for infants. The same number of babies are born, but more of them live.

There is some truth in all this, but not so much as you might suppose. Growth is curbed by providing more child care, not less. If mothers can rely on such care they learn that their babies will survive. In a poor country, your children, and especially your sons, are the only insurance you may have for your old age. But you need only so many sons. Once you have the one or two you need to take care of you when you are old, you need no more (if you are fairly confident

of keeping the ones you have) and you certainly do not want the trouble and expense of bringing up more.

Another important factor in curbing population growth, in some areas, is the increase in the educational and employment opportunities open to women. If there is an alternative to child-bearing available to women, many will postpone starting a family, and in the end the size of the completed family may be smaller. Meanwhile, the joint income of husband and wife can provide a much better standard of living than either could provide alone. They will be loath to lose that until they are sure they can provide adequately for themselves and their children, as lose it they must if the wife has a baby. If there is one word to describe the cure for excessive population growth, that word is 'hope'.

Sumatra, irrigation works. The need to drain irrigated land properly is often neglected, leading to salinization of the irrigated land. Over 200 million hectares of land are irrigated — an area four times the size of France.

Mauritius: low-volume pesticide spraying of sugar crop.

Hope of a better life is an attractive alternative to babies now. This is not to decry family life, nor to accuse people of greed. The better quality of living that can give hope is not high by Western standards.

There is another curious fact that should make us pause before we blame the victims of hunger. Ghana is a poor country. At the Sixth Special Session of the UN General Assembly in 1974, Ghana was one of the countries for which emergency relief operations were initiated. Many Ghanaians are hungry. Are there too many of them? Ghana is a country about the size of Britain, and it has a population about equal to that of London. It is far less densely populated than Britain, or than several other countries that are far richer.

The Real Cause of Hunger

No, what causes hunger is not population size, but poverty. People go hungry because they cannot afford to buy the food they need and, so far as anyone can tell, that is the only reason.

It is a paradox, but the facts support it. The world has too much food, and also vast numbers of starving people. If a commodity is in surplus, economically that has to mean there is no demand for it, or that the demand is too small. If the demand were there, the surplus would disappear. Clearly, the poor exert no demand. How can they, since they have no money?

Why are so many people poor? Is it because there are too many of them, or because they refuse to work, or because they are stupid? No, in as much as any of these statements are true – and 'stupidity' can be a consequence of malnutrition in infancy – they are symptoms of poverty, not its cause. The cause is more complex, and sadder. All of us are caught in a kind of game in which the existence of winners implies the existence of losers. It is what game theorists call a 'zero-sum' game. In it, everyone does only what seems sensible.

It works like this. In order to improve its circumstances a country needs to build a factory. To build the factory it needs money, so it goes to the rich countries to borrow the money. This is not too difficult. However, the loan is made on one condition. The plant in the factory must be supplied by firms in the donor country, and the contractors building the factory must also come from the donor country. It sounds reasonable. The donor country gains some employment and the recipient gains a factory. The loan, meanwhile, returns promptly to the country which made it.

The loan is made in the ordinary commercial way, and interest is charged on it. How shall the interest be paid? Why, from the export of the products of the factory, of course. To whom shall they be exported? To the country who lent the money, naturally. Here, however, those products meet a tariff barrier, erected to protect the internal economy. They can enter, but at a severe disadvantage. So they are not sold, the interest cannot be paid, the debts accumulate. Between 1968 and 1973, the debts owed to foreign countries grew by nearly 109 per cent in Asia, by nearly 110 per cent in Latin America, and by more than 115 per cent in Africa. Countries like Ghana are poor because so much of the wealth they create by the work they do leaves them immediately just to pay interest on loans. It cannot be used for investment at home. That is how the game traps the poor in a snare from which escape is very difficult. The hungry will remain hungry until a way out of the trap is found.

It can be found, now and then, for some countries have escaped. The fastest growing economic areas in the world at present are to be found in parts of Asia, in such countries as South Korea, Taiwan, Hong Kong, and Indonesia. In Latin America and Africa, however, help from the rich is needed, and urgently.

What Will Happen?

Conditions will improve. This is a bold, dramatic statement, but really there is nothing else that can happen because the game itself is close to breakdown. When that happens it is the rich who will suffer most.

Suppose that the economic plight of the poor countries deteriorates and their debts grow. They try to help themselves but, one by one, they fail, and default on their payments. The payments are due not to governments in the rich countries directly, but to banks in the rich countries, and it is a fact of banking that a loan is counted in the balance sheets as an asset to the bank. Consequently, their response to imminent default on a large loan is usually very accommodating. So more and more money is lent, but the debt simply grows larger until the banks themselves insist on reforms in trading patterns that will allow the poor to sell their produce to the rich at a price high enough to enable them to start reducing their indebtedness.

However optimistic this scenario, the fact remains that we could make life easier for everyone. We could cancel small debts before they cancel themselves and we could reform world trading patterns. They are, after all, man-made and not the product of some physical law. We could change the attitudes of ordinary people – ordinary rich people, that is. If the car you buy was built in your own country, then perhaps you are helping your own country by buying it. If the car was made in a poor country, very certainly you are helping the world, and without the world your own country cannot survive anyway.

Pure selfishness, then, may compel the rich to make huge concessions to the poor. As they do so they may alter the nature of the game. It may turn into a game in which no one wins huge prizes, but no one loses badly either, and everyone lives with dignity and in security. As time passes, the gap between rich and poor may narrow until finally it disappears.

If this happens, the demand for food will increase, and the rate of population growth will decline. The demand for food will rise rapidly, because eating habits will change.

At first, people who are hungry will buy the cheapest food, consisting mainly of staples such as bread or rice, with vegetables and perhaps a little meat. As they continue to grow more prosperous, this diet will begin to seem unsatisfactory. People will want a diet that contains more meat. This is what happened in Europe, so long ago we have forgotten; in Japan since 1945; it is happening now in the Soviet Union – which is why that country requires grain imports – and it is beginning to happen in those countries where prosperity is rising.

In most of the world, livestock are fed grain because the land is unsuitable for growing highly nutritious pasture grasses. Because it takes ten kilograms of plant protein, including grain protein, to make one kilogram of animal protein, an increase in the demand for meat means a very much larger increase in the demand for grain.

It is then that farmers will need to increase their output rapidly, and it is for that day they must prepare themselves. It will seem very important, then, for each country to produce as much of its own food as it can. The foreign exchange that would otherwise be spent on importing food will be needed to sustain the economic development that has increased the demand for food.

This suggests that although the world demand for food will rise, the world trade in food may not rise by the same amount. Indeed, it is possible that it will not rise at all. Even to maintain it, however, will require the stabilization of prices. For years the FAO has been urging the creation of a 'world granary' in which food might be stored in times of plenty and from which it might be released in times of shortage. Only in this way can the supply be held constant in spite of the vagaries of climate, and only by providing a reliable supply can prices be held stable.

This will be expensive, because governments will have to buy surplus food and pay for its storage, but the cost need not be unbearably high if the burden is shared among many countries. Politicians have been trying to agree on such an arrangement for some time. Eventually they will succeed.

The Ecology of Hunger

If the natural environment of our planet is being injured by the activities of humans, we

Over-rich diets literally take food from the mouths of
hungry people.

must reduce those injuries to a level that is acceptable, but we must not fall into the trap of imagining that humans and non-humans are somehow different and opposed to one another. You and I are part of the natural environment, just as much as the blue whale or the tiger that we persecute, and we have as much right to live as they.

Especially in poor countries, environmental damage is usually a direct consequence of poverty. As other contributors to this book point out, it is because they have no other fuel with which to cook their food that people fell trees and burn dung. It is because they must eat and must try to maintain a dignity that depends on the possession of herds of animals that semi-nomadic pastoralists are driven to over-exploit dry, poor pastures and so exacerbate the spread of deserts. We who are rich can well afford to treat non-humans more gently. Those who are poor may enjoy no such luxury. If we berate the slash-and-burn farmers for destroying forests, the nomads for extending the desert, the women who walk miles every day to gather fuel to cook a poor meal for their families, perhaps we are expecting of them an unreasonable sacrifice. We are, in effect, inviting them to depart this life, that the trees and grasses and flowers, the insects and birds and mammals, may live. We are being absurd. We are also being highly immoral.

The protection of non-humans must follow from the protection of humans for there is no other way in which it may be achieved. The two goals are inseparable, and environmental protection has an economic and political aspect.

Michael Allaby

Poverty is both a cause and a consequence of environmental degradation. Eight hundred million people belong to the 'absolute poor'.

Chapter 4

The Expanding Deserts

The world has always had its deserts. But never before have new ones been created as fast as now. On the fringes of the Sahara, the desert is now spreading by an estimated 1.5 million hectares (around 5,800 square miles) every year – the equivalent of creating 170 hectares of new desert every hour. In the Sudan the desert grew southwards by nearly 100 km in the seventeen years between 1958 and 1975 – a rate of advance of more than sixteen meters a day. And Africa is not the only place where the desert is advancing.

One way or another, 80 per cent of the productive land in the arid and semi-arid areas of the globe is believed to be affected by desertification – an ugly name for an ugly process. Two thirds of the world's 180-odd nations are affected by it, and the United Nations Environment Programme (UNEP) has estimated that the world's deserts – which now amount to some eight million square kilometers – could eventually treble in size. The problem concerns both developed and developing countries: 10 per cent of the United States has already been affected and a further 20 per cent is threatened.

Much of the land now being lost was once highly productive. Worldwide, about twenty million hectares of useable cropland deteriorate every year to the point where they can no longer be farmed economically. The potential value of the lost production is estimated at 26,000 million dollars a year. And even this enormous figure does not describe the total loss. Before the land finally turns into unproductive desert, its yield diminishes steadily over a number of years. With a growing world population and an increasing requirement for food – the UN Food and Agriculture Organization (FAO) believes we must increase food

MAP OF DESERTS AND AREAS SUBJECT TO DESERTIFICATION

This map shows the substantial proportion of the world's vital cropland that is vulnerable to desertification if over-used.

production by one third by the end of the century simply to maintain present levels of nutrition – it is land which mankind can ill afford to lose.

The world's area of land under cultivation is now about fourteen million square kilometers (about 5.4 million square miles). Some people believe that as much as this again has already been turned into desert. By the end of the century, UNEP has estimated, one third of the world's presently arable land may have turned into desert.

The world's natural deserts occur in

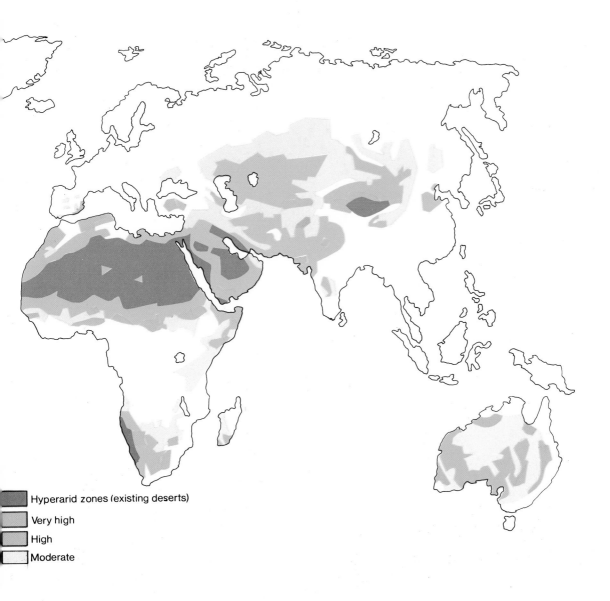

Hyperarid zones (existing deserts)

Very high

High

Moderate

areas where the average annual rainfall is less than 100 mm (2.5in). These new deserts are being created in the arid lands (rainfall less than 250 mm) and the semi-arid lands (rainfall less than 600 mm). Together the arid and semi-arid areas of the Earth account for nearly one third of its land surface, and nearly 700 million people live in them. It is widely believed that 30 million square kilometers of these lands – one fifth of the Earth's land surface – are under direct threat of desertification. The outlook for the 80 million people who live on this threatened land is bleak.

The Man-Made Desert

What causes desertification? Are deserts expanding because the climate is changing? And has the rainfall in the areas surrounding the deserts diminished? For 40 years the rather unpredictable patterns of rainfall in the Sudan have been carefully studied by scientists; the results are still uncertain. No one is quite sure whether or not there is, on average, less rain. What is certain is that the desert is growing much faster than any slight, long-term alteration in rainfall could account for. And despite the year-to-year fluctuations, the 'new desert' seems to be getting as much rain as it did in the days when it was pastureland supporting livestock. Two conclusions seem inescapable: first, if it is not the changing climate that is causing the deserts to 'expand', the new deserts must be man-made; second, it is not that the deserts are expanding but that the land near them is itself deteriorating.

In 1969 a fence was erected between the Egyptians and the Israelis in the Negev–Sinai desert. On the Egyptian side, goats, camels and sheep grazed the area intensively and Bedouins cultivated small plots of land. On the Israeli side the natural vegetation remained virtually undisturbed. Today, satellite photographs of the area show the Israeli side as dark, the Egyptian as light. The line of the fence is clearly visible. On the Egyptian side the vegetation is sparse and the sky clear; on

the Israeli side, the ground is covered in vegetation and cloud cover is far more frequent. Man is beginning to make a desert on one side, and to reclaim one on the other.

Something very similar probably happened to the Rajputana desert between Pakistan and India about 1,300 years ago. Before that time this desert was part of the Indus civilization, although today the ancient ruins lie covered in drifting sand. But strangely the 30,000-square-kilometer area is more humid than most deserts. It appears to be desert in spite of the climate rather than because of it. It seems almost certain that what happened here is similar to what is now happening along the edges of the Sahara – grazing by goats and other animals eventually removed the vegetation; the soil was bared and the microclimate altered sufficiently for the desert to take over. Climates follow deserts, it would seem, rather than the other way round.

The Negev-Sinai desert. The arrow indicates the region referred to in the text.

Certainly areas of the desert that have been fenced off and protected from grazing have developed a cover of high grasses within a few years.

Man creates deserts in many ways. The removal each year of many more trees than are planted is probably one of the most significant factors. Trees preserve the stability and fertility of the soil, retain moisture and prevent erosion by wind and water. To remove them is to invite environmental chaos. And in country after country, in Africa, Latin America and south-east Asia, local populations desperate for fuel have been stripping the land of its few remaining trees simply so that they can cook their evening meal. They are not to be blamed, for they are the victims of an international system the result of which is to keep one half of the world rich and the other poor. Desertification is one of the results of that system.

Attempts to increase food production from marginal land in semi-arid areas is a further cause of desertification. Where expanding populations of people and animals lead to overgrazing, the stripped soil is vulnerable to erosion. Modern methods of agriculture – planting single crops instead of more traditional and varied cropping cycles, and cutting down or eliminating fallow periods – leave the soil drained of nutrients and also at risk of erosion.

Irrigation schemes in areas without adequate drainage are another important cause of desertification. They can have two effects: they may raise the water-table and hence cause waterlogging of the soil; or they may bring salts and minerals to the surface, leaving the land saline and infertile. In the tropics alone, there are now thought to be at least 1.2 million square kilometers (480,000 square miles) of saline and alkaline deserts. About the same amount of land now goes out of production every year because of waterlogging, salinity or alkalinity as the amount newly irrigated. If this is not actually a losing battle, it is certainly a

Top: Kano, Nigeria: Wood for fuel is brought into the city from long distances. Centre: Jones Beach, Long Island, USA. 'Deserts' of a different kind are currently created at a rate of 8 million hectares each year by urbanization, road-building, industrial sites and other non-agricultural uses. Bottom: trees are also cut to provide forage for livestock.

static one.

The effects of overgrazing, overcropping, poor irrigation schemes and deforestation are, of course, now being accentuated by another form of desertification: the arrival of the concrete desert. Much of the world's prime agricultural land is being concreted over. Worldwide, the demand for roads and houses, factories and airports has caused the loss of between five and seven million hectares (20 to 27 thousand square miles) of arable land in the past ten years. The developed countries are, it is said, turning their best land into pavement; the developing countries are turning it into desert. As far as food production is concerned, the end result is little different.

Cash Crops that Produce Deserts

When fertile land is over-exploited – particularly in areas of low rainfall – it often turns into desert. Frequently, the traditional method of farming the land would have been shifting cultivation: a system that restores fertility by allowing land to lie fallow for two or three years after it has been cropped. The farmer would clear an area of grassland, bush or forest and cultivate it until the crop yield began to fall. Then he would abandon the site, and the natural vegetation which took over would gradually restore fertility to the soil. In some areas, a type of *Acacia* shrub was allowed to invade which could be tapped for gum arabic while the land was not being cultivated.

As populations increased, however, the pressure to raise production meant that the fallow period was cut short or abandoned altogether in favour of permanent intensive cultivation and additional grazing. The flexibility of the older systems was lost, and inevitably the over-cultivated land became less and less fertile. Nitrogen-fixing trees, such as *Acacia* and *Leucaena*, used to be part of the landscape of many countries which grew crops on semi-arid lands. Today they have been ripped out to make hoeing by

animal or machine easier. With them has gone a natural fertility which must now be replaced synthetically. And artificial fertilizer not only costs money, but usually costs money from another country, which means earning foreign exchange.

The crops grown in these marginal lands, such as sorghum and millet, were fairly drought-resistant and provided essential food. In good years a second, cash crop such as groundnuts or cotton might also be grown. Now, however, the need for foreign exchange – particularly to pay the oil bills of the OPEC countries and the fertilizer bills of the major exporters – has led to the growing of more cash crops, often on marginal land which was traditionally used as pasture.

The use of these less productive and more drought-prone areas, combined with diminishing returns from the lands under shorter fallow periods, has increased the need for artificial fertilizers still further. Eventually, of course, so much fertilizer has to be used that crop production itself becomes uneconomic. The farmers then

Soil erosion in the Sahel.

have little choice but to leave their land and seek a better life in the city. Usually they simply exchange one intolerable situation for another. But they leave behind a thin, overworked soil, robbed of the protection of its natural vegetation, and a ready prey to erosion by wind (or water, during the very heavy rains that occur from time to time even in arid areas). It is not long before these lands become desert.

Soil erosion is now a major problem in most agricultural areas of the world. Trees and hedges have been ripped out to make life easier for the tractor driver; the land is ploughed too often and too deeply; and too much reliance is placed on the kind of fertility that comes from a plastic sack. When the wind blows over this kind of exposed land, the finer particles of soil are easily swept away. This removes the most fertile part of the soil, and the part which is potentially most capable of retaining moisture.

In the United States, modern intensive agriculture has removed nearly all the natural barriers to soil erosion – the hedges, trees and ground cover. As a result, it is estimated that each hectare (ten thousand square meters, or 2.47 acres) of good cropland loses 7.4 tonnes of soil through erosion every year. Experts calculate that about one third of the arable land in the United States has been lost in the past 200 years. More and more costly, energy-intensive fertilizer has to be used just to maintain crop levels.

Erosion in developing countries is less well documented, but is certainly more serious than in the United States in the arid and semi-arid areas. The conversion of marginal land, which has survived for centuries as stable pasture land, to cropland is probably the greatest cause of desertification. The newly ploughed topsoil has a short life: what is not blown away in wind storms is washed away in sudden downpours. In Tunisia, 18,000 hectares are degraded in this way every year.

In Pakistan, 76 per cent of the total land area is now said to be affected by soil

Peru: flash floods like this carry away millions of tonnes of precious topsoil each year.

El Beshiri oasis, Sudan. Overgrazing removes vegetation cover from the soil, making it vulnerable to wind and rain.

erosion. A study in Jamaica has shown that steep slopes that are cleared of trees and used for agriculture can lose 135 tonnes of dry soil per hectare a year. The same slopes, properly terraced, lose only 17.5 tonnes a hectare. In Java, erosion and flooding are now so severe that in the Upper Solo area alone more than one thousand square kilometers can no longer be used for agriculture. It has been calculated that in one basin in this region the production of food energy will drop from 93 per cent of nutritional requirements to only 36 per cent in 40 years unless drastic action is taken to halt soil erosion.

How Bad Leads to Worse

Livestock are still grazed in the marginal lands on the edges of most deserts. The vegetation is sparse and the rains may or may not arrive, but nomadic herdsmen traditionally know the best places to take their animals, and they move on when the grazing is used up.

A useful cooperative relationship has long existed between nomadic herdsmen and sedentary farmers, who exchange and barter milk and meat for grains and legumes. Herds are allowed to graze the fallow lands in return for the fertilizing value of the dung – a practice that benefits both the livestock and the land. But newer, more intensive methods of farming are reducing the opportunities for this kind of cooperation.

As the arable farmers move insidiously into the grazing lands, less and less land is available for grazing and the herdsmen's mobility is suddenly restricted. The result, inevitably, is overgrazing. The animals strip the land of trees and grass cover. On the very arid land near the true desert, this cover is already poor. Once it has been grazed away there is no protection at all for the soil – and erosion accelerates.

One reason why desertification can occur so quickly is that overgrazing reduces the number of species of grass that can establish themselves year after

year. The more palatable grasses disappear first, because they are the first to be eaten. These are usually perennials, which play a vital role in holding the soil together. Overgrazing thus depletes the nutritional value of the grass at the same time as it destroys its ability to stabilize the land. To make matters worse, livestock tend to destroy the value of the places they like best. Around waterholes and the damper, more luxuriant areas of vegetation, the animals' hooves compact the ground, turning it into a soil pan on which nothing will grow. South of the Sahara the vegetation has been completely destroyed within 30 kilometers of some waterholes.

In rainy years, herd sizes are traditionally allowed to increase. When signs of drought appear, however,

Dinka cattle herd in the Sudan. Cattle play a central part in the lives of the many pastoralist peoples of the dry lands. As herd sizes have increased with population, so the age-old conflict between mobile herdsman and static farmer has intensified, with overgrazing as the result.

bad, it seems, then they are likely to get worse.

The Saline Desert

For many years, development experts have been trying to increase food production in the arid lands by introducing irrigation. At first sight, irrigation seems an attractive solution: it can lead to a six-fold increase in the yields of cereals and up to five-fold increases in root crops. Currently, about 250 million hectares of farm land are irrigated, some 18 per cent of all the cultivated land in the world.

Irrigation can, in theory, completely remove the danger of crop failure during droughts. And, by increasing yields, it can slow down or eliminate the expansion of cultivated lands onto the more marginal grazing land, thus helping to halt desertification. But there are problems.

Badly planned and ill-managed irrigation schemes are now a major cause of desertification. Each year, throughout the world's arid zones, 500,000 hectares of land are newly irrigated – and roughly the same area of irrigated land becomes desertified.

When dams and irrigation channels are built, the need for adequate drainage is too often neglected. Poor drainage causes waterlogging of the soil, and waterlogged soil is notoriously difficult to cultivate. To make matters worse, when the dry season comes the high temperatures encourage rapid evaporation, which brings water from lower soil levels up towards the surface. As more water evaporates, the soil becomes saltier and saltier, until a crust of salt may form on the surface. Provided that good drainage is quickly restored to such soils, the salts can be leached away and the soil reclaimed with little degradation. Soils that become alkaline, however, may in time become so compacted that they become impenetrable to the roots of trees and shrubs. They are then very difficult to reclaim.

In perhaps one tenth of irrigation schemes – usually, of course, in the most

herdsmen are naturally reluctant to decrease stock numbers, and overgrazing results. After a drought, the herd sizes are once again increased rapidly – more rapidly than the degraded land can support. Viable herd sizes and relative prosperity are, understandably enough, accorded more importance in the hungry aftermath of drought than is concern for overgrazing the land.

There is little doubt that herdsmen are now keeping too many animals for the decreasing areas of grazing land available. But populations are growing rapidly; and, as the stripped land loses its productivity, so the meat and milk output drops, resulting in a situation where ever more animals are required to keep production levels constant. This is the vicious circle of environmental decline: when things get

arid areas – a further problem occurs. If too many wells are sunk to provide the irrigation water, then the water-table may fall dramatically and the wells may dry up. Usually, the land has then to be abandoned. Even if the wells continue to work, the groundwater that is brought to the surface is likely to be saltier than the surface water. It is a perverse fact of nature that water will evaporate while salt will not. The result, inevitably, is that the groundwater and hence the soil become increasingly saline.

Iraq had depended on irrigated agriculture for 4,000 years. An irrigation project was begun there in 1953, when 60 per cent of the land was already saline.

The scheme was not carried out well: the land was poorly surveyed; the water requirements of various crops were inadequately understood; and the managers in charge of the scheme were too few and poorly trained. The drains were badly maintained and the irrigation channels silted up. By 1969 waterlogging was common, and two thirds of the soil was saline again. Yields were falling, and barley was substituted for wheat crops. Farmers had to revert to a traditional system in which they cultivated cereals under flood irrigation until the land became too saline, and then left it fallow so that the rain would leach the salts away.

In the 1960s a ten million dollar

Chad. Irrigation is the most ancient way of enhancing the productivity of the arid and semi-arid lands. About 220 million hectares of land are currently under irrigation.

kilometers of the world's best agricultural land. In 1977, the United Nations Conference on Desertification estimated that as many as 22 million hectares – nearly one tenth of the total area under irrigation – are waterlogged.

Spare a Tree – and Prevent a Desert

One of the main causes of desertification is cutting down trees. When trees and ground cover disappear, the cycling of nutrients between vegetation and soil is disrupted, and the beneficial action that tree roots have on the soil is lost. The soil is exposed to wind and water erosion, and so are the crops which were formerly protected by windbreaks.

Deforestation is creating enormous problems for those who depend on wood for cooking and heating – which means for about 90 per cent of the populations of

reclamation scheme was started. But despite attempts at drainage and repairs the canals remained silted up, the ditches filled with windblown silt and not enough drains were actually installed. In all, 64,000 hectares of land were supporting only 32,000 people, and farm income represented less than a five per cent return on the capital invested.

Much of the world's irrigated land is now believed to be prone to salinization, alkalinization and waterlogging. Salinization affects large areas of Syria, Iraq, Jordan, Haiti, Mexico, the United States and Afghanistan. Waterlogging or salinization are now believed to be seriously affecting between 2,000 and 3,000 square

developing countries. Half the total wood used, worldwide, is burned as fuel. In the tropics, 80 per cent of all wood harvested – 825 million cubic meters a year – is either simply burned as fuel or first made into charcoal.

About half of this is used in cooking, and most of the rest goes to produce hot water and space heating – although the days may be hot in the arid regions, the temperature at night can often fall below freezing point. About one sixth of the wood consumed in developing countries is used for process heat in small industries such as brick-making.

On average, each person in the developing countries uses about three

Kano, Nigeria. Cooking over open fires is still the norm in many Third World countries. Some estimates suggest that 90 per cent of all the energy used for cooking in the developing countries is wasted. Switching from an open fire to simple stoves could cut firewood use by 50 per cent.

quarters of a cubic meter of wood every year. (The average American consumes more than this in paper alone.) But in some areas of Asia and north Africa, where wood is very scarce, consumption is much less than this – simply because the wood is no longer available. The FAO has estimated that the shortage of wood in arid areas is now the equivalent of the annual production of more than 25 million hectares of fast-growing plantations. This deficit is expected to more than double by the year 2000.

In the arid tropics open woodland is now being cleared at the rate of four million hectares (more than fifteen thousand square miles) a year. Every year, Africa loses 2.7 million hectares of forest. Initially, clearing the trees may provide more land for cropping and grazing, but trees are remarkably effective at keeping desertification at bay. When they are removed the water-table may fall, and the unprotected soil become eroded, leaving the land arid and dusty. Because this is a continuing and accelerating process, statistics showing how much land is affected are hard to come by.

Just as it is possible to try to produce more crops from land than it can reasonably support, resulting eventually in depleted soil, so it is of course possible to over-exploit the forests. A certain amount of wood can be taken from forests each year; but if this amount is exceeded and too many trees are cut, the amount that is available will gradually fall – creating a vicious circle in which people are forced to burn living trees in order to keep themselves alive, thereby destroying their future source of wood. For instance, the Sahel region of West Africa is already desperately short of wood. It is believed that if the overcutting continues, only 20 per cent of the local demand for fuelwood will be able to be provided from the area by the year 2000.

For most people, wood is something that is collected rather than bought. So the shortage tends to be measured in terms of the time it takes to collect a family's needs. In Upper Volta, for example, most women have to spend between twelve and eighteen hours each week just to collect enough wood to cook their families' evening meals. In central Tanzania some households have to spend, between them, 300 working days each year to provide enough fuel for their needs. Many families in many countries now eat only one cooked meal a day simply because sufficient wood can no longer be found with which to cook two.

In the cities, where wood or charcoal cannot be collected and must, therefore, be purchased, as much as 25 per cent of a family's income frequently goes on fuel. Prices are rising steadily, and the situation encourages the greedy entrepreneur to cut live trees near at hand rather than

Ayorou, Niger. The wood market. In developing countries, as much as 25 per cent of an urban family's income may be spent on fuel.

travel greater distances to collect dead wood to sell. It is no surprise to find that many of the towns and the cities of the Sahel are surrounded by treeless wasteland.

The shortage of wood causes other problems. When people cannot find enough to cook with, they are forced to collect and burn animal dung or agricultural wastes such as straw. These, however, would normally be used to fertilize the soil. The FAO has calculated that the amount of additional food which could be produced, if all the dung that is burnt were instead returned to the land, would feed 100 million people – more than the annual increase in population in Africa, Asia and Latin America. Or, to put it another way, every tonne of dung that is burnt deprives the world of 50 kilograms of grain.

When watersheds are deforested, soil erosion leads to the deposition of silt in reservoirs that are used for irrigation, shortening their useful lives and giving further cause for alarm. And silt can ruin dams: the new Tarbela Dam in Pakistan has an estimated life of only 50 years. Some silting is expected and allowed for, but reservoirs in India are now silting up at three times the calculated rate.

Experts believe that, with enough money, large replantation schemes, sensible management and cropping of the forests, enough wood can be provided for everyone. But can it be done in time? Major fuelwood shortages already exist and more are imminent. The risk of more

India: cattle dung cakes drying for use as fuel.

and more unwitting destruction of forests is very real. Experiments have shown that, if fenced off and used only in a controlled way, depleted forest will quickly recover; but for this to be effective, there must be something there still to recover. Trees must be recognized as a valuable resource rather than as something that is simply there for the taking.

Drought in the Sahel

Desertification, although accelerating, is nothing new. But it took a major disaster to draw the world's attention to the problem.

This disaster was the drought, or series of droughts, in the Sahel region of West Africa, which lasted from 1968 to 1973. Thousands of people and millions of animals lost their lives, mainly in the six largest Sahelian countries – Senegal, Mauritania, Mali, Upper Volta, Niger and Chad. Together, these countries form an area two thirds the size of the United States.

Through most of the 1960s, until 1968, the rains in the Sahel were unusually good. The bounteous rainfall encouraged the extension of crops onto lands that had previously been considered only marginal, and better fitted to nomadic pasturing. Taking full advantage of the extra rainfall, the stock herds grew larger and their grazing was extended north to the poorer, arid lands close to the edge of the desert. These cropping and grazing intensities were far in excess of what the land could be expected to sustain on a permanent basis.

The condition of the land, therefore, was already seriously deteriorating by the time the rains failed. In 1968 there were early, heavy rains – but they stopped at the beginning of May. Before any further rain fell, in June, most of the new season's seedlings had died. There was already less land available as pasture than there had been, and now pasture growth fell dramatically. By early 1969 the first animals were dying as a result of the drought – not from lack of water but from

Up to 50 per cent of cattle were lost during the drought in the Sahel countries in the early seventies.

hunger.

In 1970, the rains failed again. The people most drastically affected were those farmers and herdsmen who, during the period of good rains, had moved furthest north. The harvest was small that year, and three million people in the region were estimated to be in need of emergency food aid.

In 1971, 1972 and 1973 the rains were again below average and the death toll began to mount. The United Nations Conference on Desertification (convened in response to the disaster in the Sahel) estimated that between 100,000 and 250,000 people died in the region as a result of the drought – and the previous over-exploitation of the land.

Goats and camels survived relatively well, but the cattle and sheep that died were counted in millions. The FAO estimated that 3.5 million cattle (about a quarter of the total) died in the Sahel in the years 1972 and 1973. Cattle losses were believed to be between 20 and 50 per cent in Mauritania and Niger, between 20 and 40 per cent in Mali and Chad, and between 10 and 20 per cent in Upper Volta. These figures do not include the animals deliberately slaughtered by herdsmen as they realized their stock would not survive the dry season. In some areas the total stock loss was, in fact, 100 per cent.

The Government of Niger released figures showing the losses in its Agadez province between 1968 and 1974 to be 88 per cent of cattle, 80 per cent of sheep, 70 per cent of goats and 45 per cent of camels.

Yet the drought was not unprecedented. There had been two other severe periods of drought earlier in the century, between 1910 and 1914, and between 1940 and 1944. The first of these lasted even longer and was even more serious than the 1968-73 drought. Obviously, occasional drought is a probability in the Sahel and, while more food needs to be grown, the region must nevertheless be prepared for further periods of little or no rain.

Zaire. Aid can often arrive too late to be of help. So it was for this refugee from Burundi, who died within hours of this picture being taken.

Meanwhile, the population in the six main countries of the Sahel is increasing by 1.4 million a year. The total population will have doubled by 2010 but, at current rates, food production will grow by only 30 per cent in the same time.

Aid to the Sahel

The Sahel disaster brought international recognition to the problem of desertification. In 1974 the UN General Assembly called for an international conference on the subject, and convened the UN Conference on Desertification (UNCOD) which was held in Nairobi in 1977, and produced a 'Plan of Action to Combat Desertification'. The UN also formed a Sudano–Sahelian Office to co-ordinate aid for the Sahel countries.

CILSS (an interstate committee for drought control in the Sahel) was formed in 1973 by Mali, Mauritania, Niger, Senegal and Upper Volta to represent the interests of the region in regard to donors of aid. Chad quickly became a member of CILSS and Gambia and Cape Verde joined later. Aid in the forms of money and food poured in: between 1975 and 1980, donors provided about seven and a half billion dollars' worth of aid to CILSS members. Per head, aid in the Sahel rose in this period from 23 dollars to 40 dollars. In 1974 CILSS declared that its objectives were now to ease the consequences of future emergencies; to make the region self-sufficient in food; and to accelerate economic and social development, especially in the least developed countries.

Unfortunately, since UNCOD little progress has been made to halt desertification. The Plan of Action to Combat Desertification contained some useful proposals, but few have been implemented. Although the UN Sudano–Sahelian Office received 162 million dollars of aid for the Sahel countries, only 8 per cent of it has been used in activities related to preventing desertification.

By 1980 aid to the eight CILSS countries was running at 1,500 million dollars a year, but only 24 per cent of this was spent on improving agriculture and forestry. And in 1975 an astonishingly tiny 0.35 per cent of all aid to the region went towards reforestation. Even in 1980, the percentage was still only 1.4.

Almost one tenth of all Sahel aid is spent on irrigation. But although this results in around 5,000 hectares of newly irrigated land coming into cultivation in the Sahel every year, a similar area of irrigated land goes out of cultivation as a result of waterlogging, salinization and alkalinization.

Five per cent of the aid is used to help livestock farmers. Additional watering points are being opened up and livestock numbers are increasing, moving steadily up towards the levels that existed before the last drought. In most Sahelian countries there are now 90 per cent as many animals as there were in 1968. In Niger there are more. Many experts believe that a further disaster is on the way.

The most urgent need now is to explore ways of increasing food production without leaving the land vulnerable to desertification – ways that will enable crops to withstand reasonable periods of drought. More money must be spent on afforestation and agricultural development. This is now particularly urgent because, since the last drought, the expansion of rain-fed cropping into more marginal lands has continued unabated, with the result that yields per hectare are still dropping. For example, the area in which cereals are grown by rain-fed cultivation in Niger has expanded by more than 40 per cent since the drought. Yet production is only four per cent higher, and the yields per hectare have fallen to an average of three quarters of their 1968 levels.

Rice production has increased since the 1960s. It is not, however, meeting demand and is subject to the whims of fashion: largely because of food aid, white bread and white rice have become popular and are replacing traditional Sahelian grains, especially in urban diets. Food aid has become a permanent fixture. Because of it, those in towns have become accustomed to imported food, and this preference is now increasing imports and decreasing self-sufficiency.

Each year more and more people in the Sahel produce less and less food. A decade after the Sahel drought, it is now clear that more is needed than the provision of aid to a calamity-stricken area. The area is still subject to drought, and desertification continues. Yet techniques are now available which could do much to stabilize this and other areas.

What Can be Done?
How can desertification be prevented? And how easily can land that has turned into desert be reclaimed?

To answer the second question first – it depends. In many areas all that is required to reclaim desert is to exclude grazing animals from it by fencing it off for a period of five years or so, and thereafter to control grazing rather strictly. The problem in areas where pressure on the land is high is that 'protection' policies of this kind can cause immense, immediate hardship. Understandably, few governments care to enforce protective legislation of that kind; and the thought that without it things may be even worse in a few years' time does little to clarify the issue.

However, protection is by no means always sufficient to reclaim a desert. Not all deserts are man-made: many, such as the Sahara itself, are the specific results of geographical or atmospheric accidents. Little can be done to make them flourish and they can certainly never be reclaimed for they have never been anything but desert – at least, not since long before Man appeared on the scene.

Even where deserts are man-made, there is now some evidence that the process is not always reversible. In southern Tunisia the tracks made by tanks and wheeled vehicles nearly 40 years ago in World War II are still visible on the ground and in the damaged vegetation. The perennial species have not re-established themselves, in spite of the fact that rainfall in this area has several times been exceptionally high and that grazing pressures are very low. This does not bode well for the future of the land now being so relentlessly invaded in the name of increased productivity.

The UNCOD Plan of Action contained 28 specific recommendations. The most important of these were that each affected nation should produce its own national plan to combat desert-ification; that regional centres be set up for research and to demonstrate what can be done, for example, to improve rain-fed cropping, irrigation techniques, livestock management and reforestation rates; and that six major transnational projects be organized in different regions which would give individual nations confidence that action was possible and inspire them to take it.

Many experts felt that the Plan was a good one which could do much to reverse the trend towards desertification. But little progress has been made. Two regional research and development centres have been established – the Sahel Institute at Bamako in Mali, and the Regional Agrometeorology and Hydrology Centre at Niamey in Niger – but few governments have produced national plans of action, and no steps appear to have been taken towards setting up any of the six transnational projects.

One reason has been shortage of money. Despite the sympathy elicited for the Sahel region, international aid for research into halting desertification has been very limited. And as each year passes, the cost of implementing the Plan of Action increases. In 1980 a study by the UN calculated that to halt desertification by the year 2000 – the major intention of the Plan – would now cost four and a half billion dollars each year. Although this figure sounds shockingly large, it must be remembered that the UN also estimated in 1980 that lost agricultural production

each year due to desertification was worth 26 billion dollars.

If the productivity of arable land in arid regions could be safely and sustain-ably increased, this would obviously help to slow down or even to stop desertification. If more food could be produced from each good hectare of land, the spread of cultivation into pasturelands could be halted – benefiting the livestock as well as the crops. Pressure would be taken off the land, reducing its vulner-ability to outside disturbance. The chances of desertification would lessen.

Irrigation is the first 'solution' that springs to mind. But irrigation is expensive and, despite all the aid money which has been devoted to it in the Sahel since the mid-1970s, only 5,000 hectares of new land have been irrigated each year. Badly designed schemes or inefficient management have caused previously irrigated land to go out of production at almost the same rate as new land has been introduced.

Despite this, there are those who argue that because of the Sahel's erratic rainfall, irrigation is essential if yields are to be maintained at high levels and if the Sahel countries are to become self-sufficient in food. However, these advocates do agree that because of past difficulties with irrigation schemes, the first priority is to rehabilitate old schemes that have now been abandoned.

About four fifths of the newly irrigated land is used to grow rice, a new crop for the area, rather than the traditional staple grains such as millet and sorghum. The newly acquired taste for rice and wheat among urban dwellers means that rice will fetch higher prices than the traditional grains, thus helping to justify the enormous cost of irrigating the land (up to 17,000 dollars per hectare). But quite how the rural population benefits – apart from the lucky few with irrigated land – is less clear.

Because of the high cost of irrigation, it seems probable that most grain will be produced in the foreseeable future by rain-fed cropping – as 95 per cent of it is at present. Research into improving yields should, therefore, concentrate mainly on this form of cultivation. A number of ideas are being considered. One urgent need is to develop high-yielding varieties of sorghum and millet that are also more drought-resistant than the varieties currently in use.

Fertilizers can greatly improve the productivity of conventional varieties; maize and millet yields have been doubled in this way. However, it is essential to use fertilizer with many improved varieties of seed, and here lies one of the major problems.

Fertilizer and improved seed are expensive items for small farmers. The yields and incomes from dry lands are currently anything but consistent and reliable, and it is difficult for farmers to raise the capital they need. Loans can sometimes be arranged but if, after a bad year, they cannot be repaid, the farmer is even worse off than before. If farmers do manage to commit funds or obtain the credit they need, they expect assurances from governments that good prices will be available for their crops. Unless governments help to arrange credit and maintain guaranteed minimum prices, many farmers understandably refuse to use fertilizer and improved seeds to improve productivity.

One advantage of increased productivity is that it might allow farmers to revert to traditional rotations, using the fallow periods, and so put the land back into good heart. A combination of the renewed use of traditional methods of caring for the land, together with the use of artificial fertilizer, could contribute towards stable and sustainable higher yields.

Additional grain storage facilities need to be provided, so that any excess grain can be stored as a buffer against hardship in years of drought and crop failure.

Some specialists believe that more livestock will have to be kept to provide enough food for the growing population.

Top: Chad. Despite its high cost, more irrigation will be essential to increase food production in the Sahelian region.

They propose to improve the quality of the grazing by first restricting growth in numbers and thus allowing the degraded lands to recover and the more valuable perennial grasses and shrubs to revive. Better livestock management could subsequently increase production without overgrazing: for example, more animals could be sold for slaughter each year; pastures could be reseeded and otherwise improved; more wells could be sunk; market routes could be better organized; and the quality of the herds could be improved by more selective breeding and better disease control.

Cattle are not very tolerant of drought, and disease control measures introduced in the past have built up their numbers only during rainy years. The cattle have died in huge numbers during droughts. Perhaps fewer cattle should be kept; there are drought-hardy breeds of sheep that have barely been introduced to the Sahel region, but that might be ideally suited to the arid conditions.

But what outside experts think is one thing and what local populations actually do is often quite another. Nomadic herdsmen regard their animals as their capital – only about nine per cent of their stock are sold in a year, largely because the herdsmen regard their animals more as a way of saving than of earning. They may exchange them with crop producers for grain but money rarely changes hands (although in recent years the same quantity of grain has come to be worth more animals). Nomads often take poorly to outside advice about what animals to breed and how to manage them.

An important way of providing a new source of fodder is to plant more trees in the marginal lands, particularly along roads and ditches and around the edges of fields. These trees not only provide extra winter fodder, but they also fix nitrogen, and hence help to improve the fertility of the surrounding land.

In attempts to improve water supply in the marginal lands, almost 7.5 million dollars have been spent since the mid-

Above: grain-storage facilities at Fadiouth, Senegal. As much as a third of all the food harvested is lost before it reaches the consumer. Inadequate storage accounts for about a fifth of this loss.

1970s on the sinking of new wells. One motive for providing new watering places is to gain some control over where, and for how long, grazing is carried on. But sinking new wells has not been successful as a measure against desertification. The wells encourage herds to remain near the water long after the local grazing has been exhausted, leading to overgrazing and the trampling of the ground over huge areas around the watering points.

Attempting to control the movements of nomads by providing wells in specific places is part of a long-term effort by governments to reduce the mobility of herds. In this way, it is believed, more control can be exerted over the factors that lead to desertification. However, the experts do not all agree that settling nomads in any of the ways proposed would necessarily be beneficial, because there is little doubt that in years of uncertain rainfall their knowledge, mobility and flexibility can help preserve livestock production.

Most attempts to improve livestock-raising in arid lands have ended in failure. This may be because planners do not have a wide enough grasp of the problems. But as a result of continual failures, planners now lack enthusiasm for dealing with the problems of livestock control and grazing, and only five per cent of development aid is now allocated to projects related to livestock.

In fact, it now seems likely that the most important way to halt desertification is simply to plant trees. Of course, planting trees under Sahelian conditions is anything but simple – though they can be grown under astonishingly arid conditions. Not surprisingly, there are few foresters in the Sahel countries and in recent years it has become clear that tree-planting is important. Once the farmers have been persuaded to accept the new plantations, they usually appreciate quite quickly the other benefits that follow: the fact that the water-table stops falling, the threat of soil erosion is reduced, crop yields are increased, there is more fodder for the

animals and, eventually, more fuelwood to burn. Because much of the land that will be needed for new forests is currently used either for cropping or grazing, such complementary planting is often the only way to secure the agreement and involvement of the farmers.

The popularity of these methods have made them the spearhead of the policies of international aid agencies such as the FAO. The trees used are usually hardy, multi-purpose species such as

A tree-planting project to reduce wind erosion by providing shelter belts.

Acacia, Leucaena and *Prosopis*, which grow quickly even on poor soils, and provide fodder as well as fuelwood and building timber. Because they fix nitrogen, they improve the soil and increase crop yields. They can also speed the fattening of cattle: some varieties of *Leucaena*, when used as part of the cattle's diet, produce weight gain at almost double the usual local rate. Research into varieties of high-yielding multi-purpose trees is now well under way in many countries.

Community – or 'social' – forestry has been hailed as one of the major answers to how to plant enough trees quickly and cheaply, to discourage desertification and to satisfy demand for fuelwood. It has certainly been a successful approach in China, which has doubled its forested area in only 30 years, planting 1.5 million hectares of new forest each year.

Although this participatory method of planting trees is only just beginning to be used in the Sahel region, a popular style

Hisurata, Libya. Dune stabilization.

has already emerged. This is to plant around a settlement area – a town or village – a belt of drought-resistant and hardy trees that will protect the settlement from winds and sandstorms, and that can be harvested for fuelwood

package: a package in which improved crops are grown, grazing lands partly protected, trees planted wherever possible, more efficient wood-burning stoves introduced and the cash-crop economy phased out.

and fodder. Ouagadougou in Upper Volta has its own 500-hectare green belt; and Niamey in Niger has a still expanding 300-hectare belt protecting it from the dry dusty winds. When the public is involved and knows that it will share in the eventual harvest from the trees, it protects the forested areas from overgrazing and fuelwood poaching; an onerous task that could never be adequately achieved by professional foresters.

Unhappily, there are no magic solutions to the problems of desertification. As in so many real life issues – as distinct from science-fiction ones – the solution will almost certainly prove to be a

The very instability of the semi-arid lands means that they will probably always be vulnerable to the slightest disruption from Man or from nature. But any serious student of desertification can quickly spot the international links in the chain of disaster which leads to the formation of man-made deserts. One of those links is the peanut, a cash crop that is not always successfully replacing traditional land uses in many semi-arid areas. It is a curious thought that the peanut-hungry West is at least partly responsible for the deserts of the South.

Robin Clarke

The Whistling Thorn is one of many species of acacia tree that can stand harsh conditions and provide both fuel and fodder.

Desertification can be stopped. The deserts can be reclaimed. What is lacking is the necessary will to do so. An expenditure of just $4.5 billion annually would halt desertification by the year 2000. That is just three days' worth of current global farming expenditure.

Chapter 5

The Mega-Extinction of Animals and Plants

The giant panda, one of the planet's most threatened species, has become the symbol of the World Wildlife Fund's global battle to preserve endangered species.

If the giant panda disappears from the face of the Earth, there will be no lack of mourners. The great beast's mournful eyes and gentle-seeming demeanour have become part of the save-species movement. Nor could we fail to note the demise of the tiger, the chimpanzee, the blue whale or the California condor. These all rank among the 1,000 or so forms of mammals and birds that, whether spectacular or bizarre or merely 'cute' in appearance, are now reckoned, according to the International Union for Conservation of Nature and Natural Resources (IUCN), to be threatened or even endangered. Mostly graceful and colourful creatures, each with its own charisma, these prominent representatives of 'wildlife in trouble' are steadily losing numbers in the wild, to the point where their plight has triggered outbursts of protests from conservationists in their millions.

But how many people have even heard of, let alone care about, the kakapo, a bird with fewer than 100 individuals remaining in New Zealand? Or the

Top: not all endangered species are pretty or cuddly. Less than 50 pairs of kakapo remain in New Zealand.

Above: excessive hunting of whales has so reduced some whale species that their populations may not recover even though they are now protected. The blue whale, one of the most threatened, is the largest animal that has ever lived.

Mauritius kestrel, now probably down to its last twenty survivors? And who would miss the Kauai Oo, fewer than ten of which hang on in Hawaii? Among the ten most endangered species in the United States are the birdwing pearly mussel, the lotus blue butterfly, the Houston toad, and the clay-loving *Phacelia*. How many of the public can even visualize these creatures, let alone would miss them? None of these organisms is as striking in appearance, or so celebrated in literature, as the tiger. But they too are threatened with extinction.

We do not hear much about these creatures with 'diminished sex appeal' because they are not included on most lists of endangered species. In point of fact, lists of endangered species contain only a trifling fraction of the world's disappearing life forms. The IUCN list of 1,000 life forms is almost entirely made up of mammals and birds, these being the most popular and most readily recognized of Earth's wild species. Actually, IUCN has recently drawn attention to the 25,000, or one in ten, of Earth's plants that are threatened; and an attempt has been made to do the same with vertebrates other than mammals and birds, namely the reptiles, amphibians and fishes. But all vertebrates put together total only a little over 40,000 species, and plants no more than one quarter of a million. All together we share the One-Earth home with between five and ten million species, the great bulk of them being insects – and it is among these creepy-crawly creatures that

The tiger was the first animal for which an international rescue effort was launched. Tiger populations in India are now increasing.

the greatest numbers of extinctions are occurring. In fact there are between 100 and 200 times more species of insects than there are of vertebrates. Equally important, scientists estimate that they have identified less than one fifth of all insect species, let alone assessed them for their survival prospects. If the status of all these myriad creatures were examined, we would find that hundreds of thousands of their species are teetering on the abyss.

Extinction Rates

In short, we have to recognize the consequences of our ignorance. Animal forms that have been documented and recognized as threatened now amount to over 1,000. Yet even though this is a shockingly large number, it represents only a tiny fraction of the problem. Whereas these 1,000 life forms are probably disappearing at a rate of about one per year, species overall are probably disappearing at a rate of at least one per day. By the end of the 1980s we may well be losing one species per hour – unrecorded and apparently unregretted. By the end of this century we shall be fortunate if we have not said goodbye to one million of the 5–10 million species that make up the planetary spectrum of

Zoos and circuses have introduced chimpanzees to millions of people. But have they also endangered their existence?

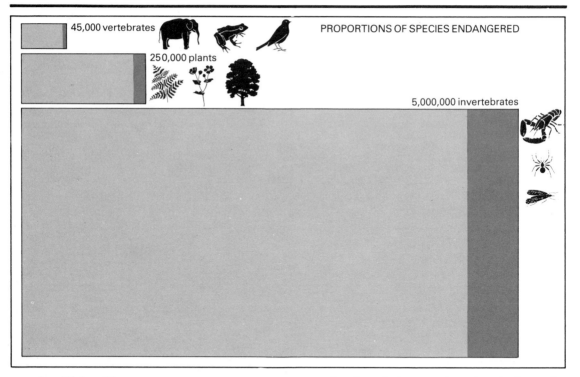

45,000 vertebrates

250,000 plants

PROPORTIONS OF SPECIES ENDANGERED

5,000,000 invertebrates

species. This will represent a greater biological debacle, in terms of its magnitude and compressed time-scale, than any extinction spasm since the first flickerings of life 3.6 billion years ago. When the dinosaurs and their kin, plus many marine creatures, toppled over the edge into the void of extinction 65 million years ago, the mass extinction was so large and so swift that geologists refer to it as 'the great dying'. Yet these species disappeared at a rate of no more than one every 1,000 years at most.

Thus the present phenomenon is unprecedented by any measure. In a twinkling of an evolutionary eye, we are witnessing the elimination of a sizable sector of life's diversity on Earth. We are not merely making pinprick holes in the fabric of life: we are imposing vast rents upon it.

Causes of Extinction
Virtually all these extinctions are being caused through the indirect actions of humankind. We very rarely hunt a species

to extinction. It is true that the dodo and the Steller's sea-cow, the passenger pigeon and the great auk, were driven over the edge through crass over-hunting as much as by any other factor. In our own age, we have seen a similar fate all but overtake the great whales (fortunately, the whales are now entering a more prosperous era, having been rescued virtually from the brink of extinction). But in the main, wild creatures are not eliminated through the poisoned arrows, the harpoons and the sporting rifles of modern man. Rather they are eliminated through the bulldozer, chainsaw and digging hoe. In Africa, comparatively little damage is being inflicted by the many thousands of spears and traps utilized by men who know they are breaking the law. Much more harmful are the millions of digging hoes, wielded by people who are otherwise landless, and who believe they are doing no more than pursue their legitimate needs as law-abiding citizens when they dig up the last habitats of zebras, giraffes and lions.

Thus, a spasm of extinction is overtaking the diversity of life on Earth. Eventually we can hope that humankind will stabilize the growth of human numbers, and will regulate the growth of material appetites, until we strike some sort of ecological accord with our living space. But by that time it is probable that we shall have lost one quarter of the Earth's species. It is possible that we shall have lost one third. It is not inconceivable that we shall have lost one half.

Tropical Forests
Some readers may well be asking whether there is much evidence for the grim scenarios outlined above. Well, let us look at one particular zone, the tropical forests. These forests cover only seven per cent of Earth's land surface, yet they harbour between 40 and 50 per cent of all species. Equally to the point, these forests are being disrupted and degraded, if not destroyed outright, more rapidly than any other ecological zone. If present over-exploitation persists – and it is likely to accelerate – many tropical forests are not likely to survive except in severely disrupted form by the end of the century, and their remnants will have poor prospects for long-term survival of any worthwhile sort.

'Dead as a dodo' has become a cliché which tells its own sad story. The dodo, a flightless bird, is the most famous of all the species that have become extinct as a result of man's intervention. A native of the Pacific, its inability to fly made it an easy prey for passing mariners.

Unlike temperate-forest species, which are far less abundant, tropical-forest species tend to have specialized ecological requirements. They may be limited to habitats totalling a mere few hundred square miles. Or they may occur at very low densities, only one individual in every dozen square miles at most. These attributes leave them unusually susceptible to summary extinction when their habitats are invaded by modern man with his disruptive activities. By dint of axe, chainsaw and bulldozer, many millions of people are eliminating tropical forests in West and East Africa, Central America and the fringes of Amazonia, and, most of all, southern and south-east Asia. Not that the story is entirely dismal: because of their very low population pressures, and their remoteness, we can expect that much of the Zaire Basin and the Amazon Basin will survive little harmed until well into the next century. But by the year 2000, we should not be surprised if we lose between one third and one half of the tropical forests that we had in 1950. With their disappearance, we shall witness the mass extinction of hundreds of thousands of species.

The first wave of these extinctions is already occurring in south-east Asia's rain forests, some of the wettest and richest forests of the entire tropical forest zone. Just as forests represent a biological treasure-house to the scientist, so they represent a commercial gold-mine to the timber harvester. By the time you have read to this point in this chapter, you can expect that many dozens of forest giants have been felled in south-east Asia, some of them almost 200 feet high and hundreds of years old, yielding several tonnes of specialist hardwood timber that is in great demand by rich and poor countries alike – but especially by rich countries, with their penchant for parquet floors, fine furniture, fancy panelling, weekend yachts, and high-grade coffins.

Not only are tropical forests the sites of accelerating extinctions. Similar processes of gross disruption are

Caribbean

Central America

South America

Although accounting for only seven per cent of the Earth's land surface, the tropical forests harbour about half of all the species on the planet. Many, especially insects and plants, are not yet known to science. About one per cent of the forest is destroyed each year.

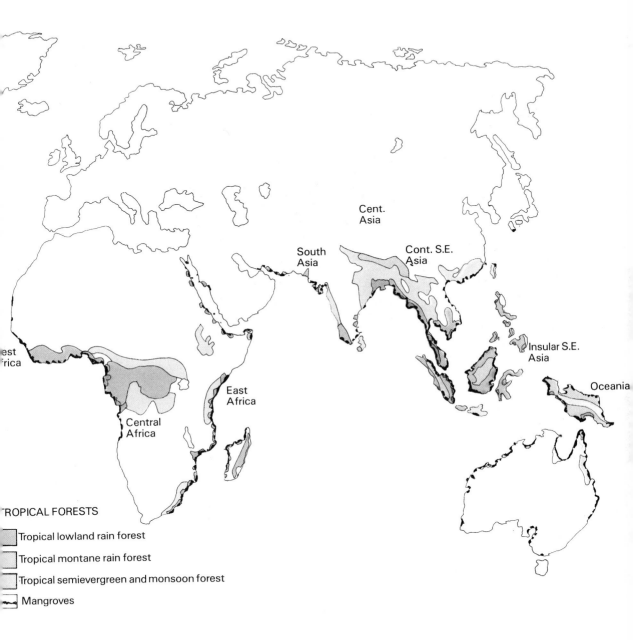

Cent.
Asia

South
Asia

Cont. S.E.
Asia

Insular S.E.
Asia

Oceania

est
rica

East
Africa

Central
Africa

ROPICAL FORESTS

Tropical lowland rain forest

Tropical montane rain forest

Tropical semievergreen and monsoon forest

Mangroves

overtaking other ecological zones that feature exceptional concentrations of species, notably coral reef ecosystems and wetlands.

If tropical forests alone contain between two and four million species, and other tropical ecosystems as many again, it is not difficult to envisage that, by the year 2000, we shall witness the demise of one million species. This will work out, during the course of the last twenty years of the century, to an average extinction rate of 50,000 species per year, or 137 species a day. The maximum exploitation pressures will not come to bear on tropical forests and other species-rich biomes until the latter part of the period. That is to say, the 1990s could see many more species wiped out than the previous several decades put together. But already the destructive processes are gathering momentum. Hence it is all too realistic to suppose that we are already losing one species per day.

Above: Malaysia's forests are among the wettest and richest of the tropical forests. The insatiable appetite of the rich countries for hardwoods means that by 1990 most of Malaysia's once lush forests will have been cut down.

Top: wherever man goes, rats, sparrows, cockroaches and other hardy species follow. As human beings transform and simplify natural ecosystems so they eliminate specialized species which depend upon very precise habitats. This creates room for those species which are most adaptable — of which rats are one of the most successful — to rush in and proliferate.

Many of our fourteen crane species, including the whooping cranes shown here, are threatened by hunting and habitat destruction. Vigorous efforts have been made to protect this species from extinction, but their long-term success is by no means assured.

Speciation

Of course, when large numbers of species disappear, they open up a lot of 'ecological living space' that can be occupied by other, newer species. In other words, an outburst of extinctions may lead to an outburst of speciation, the process by which new species come into being. Regrettably a new species emerges excruciatingly slowly as compared with the great speed of man-caused extinction. Whereas speciation often takes hundreds of years, generally thousands of years, and sometimes millions of years, extinction can occur in less than one decade, occasionally in just a few months. Nonetheless the foreseeable future may witness a speeding-up of certain aspects of evolution. As large slices of life's array disappear, new 'niches' will open up for newly-emerging species to occupy – and as multitudes of niches become vacant, the phenomenon might well stimulate a spurt of speciation.

Unfortunately, the first communities of such species to emerge tend to be pioneering types. They are opportunist creatures, adapted to rush in and proliferate with speed to colonize a new territory overnight. Examples of these 'clever' species are rats, sparrows, cockroaches, sundry insect pests, and 'weedy' plants. Through unconsidered intervention in evolution's course, we may be triggering processes that will populate the world of our children with 'plague species' of multiple kinds – a prospect made all the more dismal if, in the meantime, we have eliminated those species that, as predators and parasites, could keep the opportunist species in check. In fact it is all too likely that these 'natural enemies' of pest species will suffer disproportionate extinction.

Thus we have the problem of disappearing species – one of the great 'sleeper issues' of our time. It is difficult to imagine a problem more profound in its implications, yet less appreciated by the public at large. During the past few years, increasingly alarmed noises have been heard about a few prominent species such as the chimpanzee, the cheetah and the polar bear. Through exceptional efforts, several of these endangered animals have been brought back a step or two from the brink; the whooping crane probably numbers twice as many individuals as a dozen years ago. Yet during the course of these same few years, it is likely that several thousand species have been pushed over the edge into oblivion.

What is in It for Us?

Extinction of species represents an irreversible loss of unique natural resources, now and for ever. Our planet is currently afflicted with various other forms of environmental degradation, notably pollution. But if we decide to clean up our skies and rivers, we can generally do so: it will cost us a great deal, but the process is essentially reversible. Not so with extinction. When a species disappears, it is gone for good.

Often enough, that will be for bad. People around the world now consume many foods, take numerous medicines and utilize a myriad of industrial products that owe their origin to wild species. Of the one per cent of all species that have been intensively investigated for their economic value, many thousands already make sizable contributions. In light of experiences to date, it is a statistical certainty that Earth's species potentially offer many utilitarian benefits to society.

In fact, wild species rank among the most valuable raw materials with which society can meet the unknown challenges of the future. Yet the Earth's stock of them is being depleted more rapidly than many of the planet's most precious mineral deposits.

Medical Benefits

When we take a medical prescription to a pharmacy, for example, there is one chance in two that the medication we collect owes its existence to materials of 'natural' origin. The commercial value of naturally derived drugs in the United

States is now valued at over ten billion dollars per year. Sales of non-prescription preparations that similarly derive from wild creatures are worth another ten billion dollars.

If forced to give up a cardiotonic compound called digoxin, produced from the Grecian foxglove, more than three million Americans with high blood pressure would find their lives cut short within 72 hours.

A Caribbean sponge has been found to contain a compound effective against herpes encephalitis, a deadly brain infection that strikes many thousands of people each year and against which there has previously been no worthwhile drug. The sponge's compound has supplied a breakthrough in the treatment of diseases caused by viruses, much as penicillin did for diseases caused by bacteria. As a result of this widely hailed discovery, there is now the prospect of curing a wide range of viral diseases, from the common cold upwards.

Digoxin is a widely used drug for controlling heart disease. It is one of a family of useful chemical compounds called alkaloids found naturally in plants. Digoxin is produced from the familiar foxglove.

In 1960, a child contracting leukemia had one chance in five of survival; since then scientists have developed a drug – vincristine – from a plant of the tropical forests, the rosy periwinkle, that now allows a leukemia sufferer four chances in five of survival. The National Cancer Institute near Washington DC has screened 29,000 plant species for potential use against cancer. About 3,000 show preliminary promise, and at least five may come to rival vincristine. The Institute believes that mass extinctions of species could represent a serious setback to the future of anti-cancer campaigns.

Among other medical products, the 'pill' that is swallowed by 80 million women each day contains sex-hormone combinations derived from a Mexican forest yam. Over-the-counter sales of the pill are now worth almost one million dollars a year. The World Health Organization is conducting a worldwide search for materials with which to manufacture a safer and more effective contraceptive pill – and the agency considers that its best bet lies with tribal peoples in tropical forests who have traditionally used any or some of 3,000 plant species for their anti-fertility properties.

As for more esoteric applications of wild species, the lobelia plant yields organic alkaline compounds, known as

94

This pretty plant is a lifesaver, literally. Until 20 years ago leukemia was a certain killer. The scientists discovered that the plant, the rosy periwinkle, produced a drug which greatly enhanced the survival chances of leukemia sufferers.

alkaloids, used in anti-smoking preparations. New plant sources may soon be found for codeine and to replace the morphine that also produces heroin. A substitute source that would serve licit medical needs without fostering drug abuse would relieve expensive efforts to combat one of our greatest social problems.

During the course of 1983, supposing that we are losing one species per day, we may have already said goodbye to more than 260 species of plants and animals by the publication day of this book. The sun has continued to come up, and the world to turn, without the slightest palpable effect on the average citizen going about his or her daily affairs. Yet who knows whether those 260 species included a rosy periwinkle, a yam or a lobelia that could serve our welfare in ways we have yet to imagine?

Agricultural Benefits

Much the same applies to our daily diets. Almost one third of Earth's 250,000 plant species are believed to be edible, and humans have used at least 3,000 for food at one time or another. So why do we stick to fewer than twenty species for 90 per cent of the food consumed around the world each day? The US National Academy of Sciences, in a 1974 survey, came up with several hundred species

Wild grain. Farmers, and their scientist allies, are ingenious. So, too, are the pests against which they wage constant war. As new insect pests emerge, or old ones adapt, so new strains of plants resistant to their assaults must be developed. To do this, plant geneticists must be able to draw on fresh genes from wild varieties of grains and other crops. As we bring more and more virgin lands under cultivation or management, so we reduce the pool of genes from which to breed new varieties of crops.

that appear to offer immediate potential to relieve hunger, improve nutrition and offer tastier menus. A marine plant from the west coast of Mexico produces grain that can be ground for flour, opening up the prospect of using the seas to grow bread. Also found in Mexico are various insects which, properly cooked, are said to rank as delicacies.

Apart from new foods, wild species help established agriculture. Corn, wheat, soybeans – all conventional crops – need to have their genetic constitutions regularly 'topped up' in order to maintain and expand their productivity, and in order to resist new insect pests, changes in weather patterns, or similar environmental threats. The fresh germplasm comes from wild varieties and primitive cultivars of crop plant species.

It is the skills of plant geneticists, rather than fertilizers and pesticides, that have led to one record after another in crop yields in North America, the 'green revolution' countries of the tropics, and in other parts of the world. The United States and Canada are especially dependent on exotic gene reservoirs; their agriculture is essentially an imported agriculture. All their main food crops having originated in Latin America and elsewhere, they must rely almost entirely on foreign supplies for genetic resources. The US Department of Agriculture estimates that annual increases in crop productivity due to genetic improvements are worth at least one billion dollars.

This amount will be greatly increased if geneticists can exploit a perennial form of wheat recently discovered in Mexico's forests. Hybridized with existing varieties of wheat, it could eliminate the season-by-season ploughing and sowing that now cost the farmer (and thus the consumer) huge sums each year.

Agriculture can also be boosted through reduction of the damage caused by insect pests, which costs us five billion dollars per year. A number of pests can be controlled by other insects, notably predators and parasites. Since many of these control species are highly specific in their choice of prey, they limit their attentions to target species of insect pests without doing damage to other harmless species. This strategy is far preferable to broadscale use of persistent toxic chemicals, to which many insect species are growing resistant. As many as half of the 800-plus pests that damage crops around the world fall in the latter category.

There are now more than 250 cases of partial or complete control of insect pests (also weed problems) through the introduction of insect predators and parasites. These 'counter-pest' species are estimated, according to the Food and Agriculture Organization of the United Nations, to return 30 dollars for every one dollar invested.

Industrial Benefits

As for industry, plants already serve the needs of the textile manufacturer, the toilet-goods producer and the ice-cream maker – likewise the butcher, the baker and the candlestick-maker. As technology advances, in a world growing short of just about everything except shortages, industry's need for new raw materials will grow ever more rapidly.

The prospects are obvious when looking at just one category: fuel. As Nobel Prize-winning Melvin Calvin of Berkeley, California, points out, certain plant species produce hydrocarbons, like oil, in place of carbohydrates, like sugar. These hydrocarbons can be of various kinds, one of which – rubber – we have long used and which comes from a tree of the *Euphorbia* family (a tree that, 100 years ago, was considered a 'nuisance plant' with no apparent use to Man, upsetting the exploitation plans of tropical foresters). Several other *Euphorbias* produce significant amounts of a milk-like sap – latex – that is actually an emulsion of hydrocarbons in water. These hydrocarbons are superior to those of crude oil, since they are practically free of the sulphur and contaminants found in fossil petroleum.

Altogether some 38,000 species of plants produce latex, but the *Euphorbias* seem to be especially suited to 'growing gasoline' – notably twelve species from Brazil that can be grown in areas too dry for other conventional purposes, such as agriculture, or on land that is otherwise useless, such as strip-mined areas.

According to Calvin's experiments, one acre of *Euphorbia* trees could produce between 10 and 50 barrels of oil per year, at annual production costs that now run at about 25 to 30 dollars per barrel, compared with OPEC prices at almost the same level. Geneticists have no doubt that they can double production through seed selection within a year or two, while agronomists believe they can achieve similar increases in output within a few years.

After all, the rubber tree in 1945 was producing a mere 200 pounds of rubber per acre, an amount which scientists pushed upwards ten-fold within the space of 20 years. A few trees were even bred that yielded 40 times as much. 'Petroleum plantations' on a commercial scale have been established on Okinawa by two large Japanese corporations, Nippon Oil and Sekisui Plastics.

What are We Doing?

So much for the problem. What is the response of the scientists, conservationists, economists, politicians and others that contribute to the husbandry of our Earth?

Developing-world leaders are now becoming aware that they possess the

Plant geneticists studying the effects of the semi-dwarfing gene RHT in different wheat backgrounds.

great bulk of the Earth's species and their genetic resources. At the same time, it is the developed nations, technologically advanced while biologically impoverished (by comparison with the tropics) that can best exploit the germplasm reservoirs of the Third World. As we have seen, a number of plants with anti-cancer properties are being discovered in tropical forests; and the countries in question are realizing that the benefits will accrue, for the foreseeable future at least, to so-called advanced societies of the temperate zones – for tropical-zone citizens do not generally live long enough to contract cancer. To the extent that the developing nations of the tropics are trying to preserve their species, their efforts amount to a 'resource handout' to developed nations, a situation that Third World leaders increasingly protest about, and wish to see raised within the context of North-South negotiations. As one Third World personality has put it, 'How can a developed-world conservationist speak of the International Union for Conservation of Nature without speaking at the same time of the New International Economic Order?' This approach contrasts with the conventional view of the traditional conservationists of the rich nations, who believe that the extinction of species represents an impoverishment for the entire global community, in so far as species constitute part of everybody's natural heritage. Developing-nation leaders, however, are inclined to perceive the problem, with ever-growing urgency, as part of the 'resource confrontation' between the North and the South.

To resolve these two different approaches, two initiatives could help. The first would be an explicit acknowledgment by the community of nations that, whereas species represent an indivisible part of humankind's patrimony, a disproportionate share of the preservation burden now falls on those nations least able to bear it. A second initiative could lie with a proposal that some observers consider realistic (and

others idealistic): it suggests that rich nations might consider a cost-sharing arrangement to assist developing nations with the expense of preservation campaigns in the tropics.

Without initiatives along these lines, species may increasingly be regarded as a further set of high-value resources through which the South expresses its grievances against the North. Thus, at any rate, rumblings in the corridors at recent meetings of United Nations agencies have suggested.

It is true that there are already some tentative steps along these lines, such as UNESCO's World Heritage Trust and its Biosphere Reserves project. Both these initiatives are, however, absurdly under-funded. As a minimum requirement, each of Earth's bio-geographical provinces (there are almost 200 of them) should

The Euphorbia family of trees — which includes the rubber tree — can produce a sap containing complex hydrocarbons. In the future we may be able to 'grow' gasoline.

feature at least one Biosphere Reserve; to date, fewer than one quarter of these provinces possess such a Reserve. To establish a typical Reserve costs around 100,000 dollars, and to maintain it another 50,000 dollars per year; so a project to set up another 150 Reserves, and to run them until the end of the century, would cost 165 million dollars – the equivalent of a value-added tax of 0.1 per cent on internationally traded oil (as has been proposed by Saudi Arabia) extended over a mere twenty months. Kindred sources of revenues could lie with a value-added tax on international trade of all kinds, this measure impinging most directly on those sectors of the global community that benefit most from exploitation of the Earth's natural resources – the developed nations. A trifling 0.1 per cent tax on such trade would yield one billion dollars a year.

Pragmatic measures of this kind would help to put operational muscle behind the newly launched World Conservation Strategy, released in early March 1980 by the International Union for Conservation of Nature and Natural Resources (IUCN). A major part of the strategy focuses on threatened species. The conceptual analysis of the strategy is first-rate, but the strategy does not spell out clearly enough how we get from here to there. The main problem lies with funding, and with the cost–benefit arithmetic of conservation activities that involve all nations of the world and that will benefit all generations into the future. However warm-hearted one may feel about threatened creatures, one must be hard-nosed in devising the institutional mechanisms, the political initiatives and the economic rationale, to enable a conservation strategy to survive in our workaday world.

A 'Triage Strategy' for Threatened Species

Until such time as sufficient funding becomes available, we shall have to do our best with the limited conservation resources we can bring to bear on the problem. It is becoming clear that even if present funding were quadrupled, we could not save all those species that appear doomed to disappear. The processes of habitat disruption are too solidly under way to be halted completely. Since Man is intervening in the

evolutionary process with as strong an impact as (for example) an Ice Age, he should do it with as much awareness of what he is about as he can muster. That is to say, he has failed at playing Noah, in so far as the Ark that he has designed is far too small for the job. So Man is now committed to playing God instead, on the grounds that he is determining extinction of large numbers of species. In these circumstances, he might play his role with as much selective discretion as possible. But how to accomplish this? If we cannot be sure of the details, can we at least establish the right direction to move in?

MAP OF PRIORITY AREAS FOR PROTECTION

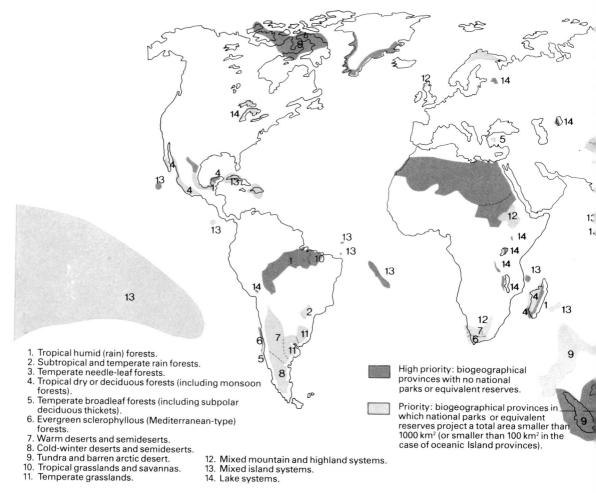

1. Tropical humid (rain) forests.
2. Subtropical and temperate rain forests.
3. Temperate needle-leaf forests.
4. Tropical dry or deciduous forests (including monsoon forests).
5. Temperate broadleaf forests (including subpolar deciduous thickets).
6. Evergreen sclerophyllous (Mediterranean-type) forests.
7. Warm deserts and semideserts.
8. Cold-winter deserts and semideserts.
9. Tundra and barren arctic desert.
10. Tropical grasslands and savannas.
11. Temperate grasslands.
12. Mixed mountain and highland systems.
13. Mixed island systems.
14. Lake systems.

High priority: biogeographical provinces with no national parks or equivalent reserves.

Priority: biogeographical provinces in which national parks or equivalent reserves project a total area smaller than 1000 km² (or smaller than 100 km² in the case of oceanic Island provinces).

Most of the areas needing priority attention for protection lie in the developing countries. They will need financial help from the rich countries if they are to establish the necessary reserves.

These are large questions to ask. How are we to decide which species shall be allowed to become extinct through our deliberate decision, and thereby concentrate our conservation efforts – limited as they are bound to be – on more 'deserving' species? This would mean that certain species would simply disappear because we pulled the carpet out from under them. We might abandon the Mauritius kestrel to its all-but-inevitable fate and utilize the funds to proffer stronger support to any of the hundreds of threatened bird species that are more likely to survive. In short, a proportion of species would disappear through human design. Agonizing as this prospect might be, it is better than allowing species to disappear merely through human default.

An approach along these lines would amount to a 'triage strategy' for species. The term derives from French medical practice in World War I, when battlefield doctors found there were more wounded than they could handle. So they assigned each soldier to one of three categories: first, those who would certainly be helped by medical attention; second, those who could probably survive without attention; and third, those who were likely to die no matter how much attention they received. The first category absorbed pretty well all the medical services available, so the other two categories were ignored. If a strategy along similar lines were applied to threatened species, it would amount to a more rational approach than that practised hitherto. It would be systematic rather than haphazard, and it would enable conservationists to make the best use of their finances and skills.

How will choices be made? How shall we decide between the Bengal tiger and a crab in the Caribbean? These will be difficult decisions. A start could be made through methodical analysis of the factors that make some species more susceptible to extinction than others; for example, sensitivity to habitat disruption and poor reproductive capacity. In addition to bio-ecological factors, there is a need to consider the social, economic and political aspects of the problem: the Bengal tiger requires large amounts of living space in a part of the world that is crowded with humans, but it could stimulate more public support for conservation of its ecosystem (and thereby help save many other species in the same ecosystem) than could a less-than-charismatic creature such as a crab. Certain species would appear to be solid candidates for consideration by virtue of their utilitarian value. For example, the one plant species in five that contains organic alkaline

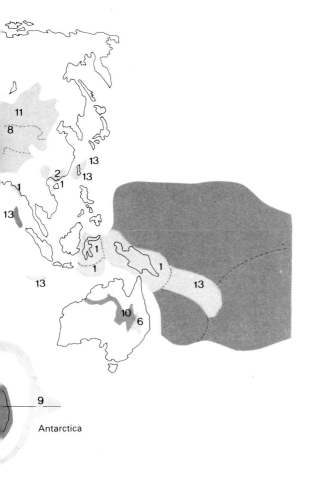

Antarctica

compounds, known as alkaloids, is more likely to produce new medicines than other plant species. The one plant species in twenty that is a legume may, in so far as it fixes nitrogen from the atmosphere, prove a source of fertilizer, thus supplying support for agriculture. The *Ichneumonidae* family of insects, containing wasps among others, have shown themselves exceptionally capable in controlling insect pests. The few tree species, notably *Euphorbias*, that produce hydrocarbons in their tissues, instead of carbohydrates like sugar after the manner of most plants, open up the prospect of 'growing gasoline'.

Many tough decisions will have to be made. Nobody will like the challenge of deliberately consigning large numbers of species to oblivion. But in so far as Man is haphazardly consigning huge numbers to oblivion already, he might as well do it with more systematic judgment than he has been able to apply so far.

The California Condor

A case in point is the California condor. With a spectacular wing spread of over three meters, the bird can soar to 3,000 meters, and live to be 35 years old. A majestic sight (on the few occasions when it reveals itself to the human eye), the condor excites a curious mystique in the public's mind. Presumably this is due not only to the condor's super-spectacular appearance, but to its plight as one of the most endangered bird species in the United States, if not on Earth. Hence the symptomatic appeal of the condor.

Although the largest bird in North

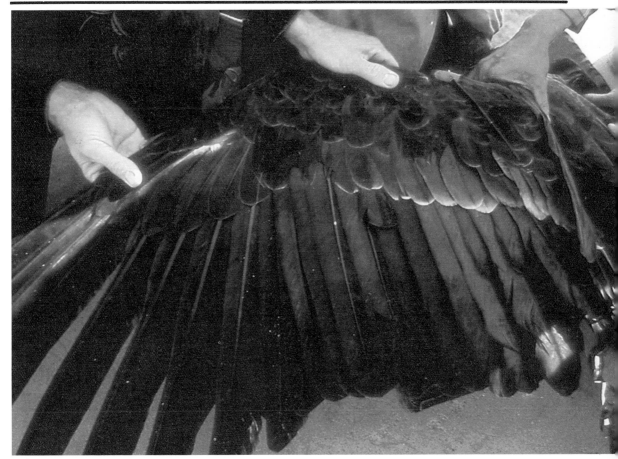

America, the condor has one of the smallest populations. By 1980, the species' total had declined to around 30 individuals or so, a mere eight or ten of which are of breeding age. (There is also one captive bird in the Los Angeles Zoo.) Despite complete protection for the birds, and the isolation of their nesting areas in high-cliff caves, the species apparently pursues a slow and inexorable march towards extinction.

How far should we go to protect a creature that is dismissed by some observers as an evolutionary 'geriatric case'? Should we assume that we do not yet know how to determine when a species is inevitably declining through its own natural accord, and try to save the creature? This option is currently being pursued, at a cost of some 25 million dollars spread over forty years. The preservation programme, one of the most expensive conservation attempts ever undertaken, offers, according to the US Fish and Wildlife Service, only a fifty-fifty chance of success – and we may not know whether the investment is paying off for at least another two decades.

Despite these reservations, however, the effort may represent a sound use of conservation funds, in view of the condor's symbolic value. Suppose the bird were allowed to slip quietly into oblivion, an extinction that could be construed as part natural, part Man-caused. Might the public not protest that if conservationists cannot save the condor, what can they save? In other words, does the bird possess 'public opinion' value way beyond the value of its own intrinsic worth? If this

103

In October 1982 the Condor Research Centre trapped an immature male condor. A small solar-powered radio transmitter was fixed to each wing and the bird released. The transmitters allow the birds to be tracked over distances of up to 100 miles.

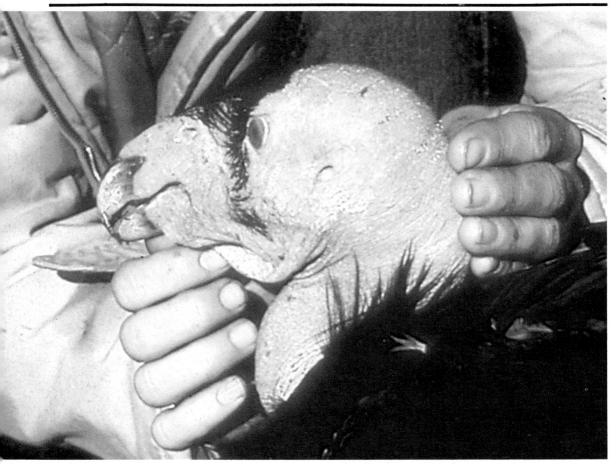

is the case, then we should go ahead with a sizable outlay of conservation funds in support of this single species, despite its doubtful prospects of survival.

The decisions about the condor are difficult in the extreme, largely because the values at stake cannot be quantified. So, at least, the argument runs. But one exercise in quantification could illuminate the issue – and, to this writer's knowledge, it has not yet been undertaken. An allocation of 25 million dollars to the condor means that a similar sum cannot be spent elsewhere. If the same monies were given over to a preservation plan for one of Hawaii's richer habitats, they might help the cause of several bird species at one time, all of them offering far better chances of success than the condor (and a Hawaiian project could protect a number

of other rare and threatened species, notably plants and insects, at the same time, by contrast with the situation in the condor's habitat). Still more to the point in terms of the global heritage of species, 25 million dollars spent on the relict montane forest of East Africa, or on the southern strip of Atlantic-coast forest in Brazil, or on certain localities in New Caledonia, would almost certainly assist several dozen threatened species – none of which, however, possesses the charismatic image of the condor.

Ecosystems of Exceptional Value
As this chapter has made plain, certain zones are biotically richer, and ecologically more diverse, than others. Notable examples are tropical forests and coral reefs. By safeguarding a sector of

The Californian Condor Research Centre now has a busy research programme which is not only investigating the foraging, breeding and other ecological relationships of the condor, but is also studying the impact of human activities on its well-being.

these biomes, conservationists can accomplish more, in terms of saving totals of species, than through safeguarding much larger zones in other biomes.

A similar approach can apply to other ecological tracts. The Sudd Swamp in southern Sudan, being a long-established island of moisture amid a semi-arid region, features an exceptionally rich array of species, many of them endemic. The Sudd's ecosystem is currently being disrupted by the Jonglei Canal, and entire communities of species could become extinct within the foreseeable future. Many other wetland zones can be identified, deserving varying degrees of priority treatment for conservation.

To this extent, then, conservationists can avoid the dilemmas of ranking species by priority, by directing greater attention to the protection of entire communities of species, and the protection of entire ecosystems. (This expanded approach is already proclaimed by conservationists; but it is observed more in principle than in practice, since the great bulk of efforts of, for example, the Species Survival Commission are still directed at individual species, rather than communities or ecosystems.) Yet even when we direct our conservation towards the broader level of communities and ecosystems, we are still faced with the same agonizing choices: how do we choose between those communities and ecosystems that would be very important to save, and those that are *essential* to save – given that we cannot assist the whole lot, due to lack of funds? How would we prepare a hierarchical ranking among, say, tropical forests, coral reefs, and tropical wetlands? This would be a difficult decision indeed.

The choice need not necessarily be presented to us in this perplexing form. Were all conservation resources to be directed at these three ecological zones, as clear priorities ahead of the rest of the field, we would probably not have to make choices between them, since our resources would then be sufficient to do a good job on each of the three categories.

Alas, that is not the way the conservation world works, and the great bulk of conservation funds, originating in the rich nations of the temperate zones, continue to be directed at species and habitats in temperate zones, even though this is where the problem of threatened species is not so acute, nor suffering such a dire lack of funds, as in the tropical zone of the Third World.

The Need to Choose

Let us recognize the urgent necessity of making choices among threatened species. This is not a formidable challenge to be confronted somewhere down the road in the future, when we have had a chance to think things out. It is a challenge that we already face right now with the funding-allocation systems that we already employ – less than systematic and rational as some of those approaches are. Ever since the start of the save-species movement, we have been making choices between species. The expanded strategy proposed here amounts to no more than an extension of the past, albeit in a more methodical manner. The key question is not, 'Shall we now attempt to apply triage?' The question is, 'How shall we apply triage to better effect?'

Regrettably, the term 'triage' tends to raise negative connotations in the minds of some observers. Yet a triage strategy applied to threatened species would amount, in many respects, to a better approach than that which has generally been practised hitherto. It would be systematic rather than haphazard, and it would help conservationists to make optimal use of their finances and professional skills. Those threatened species that, for biological or economic or socio-cultural reasons, present 'the most productive opportunities' for investment of conservation resources, should clearly come top of our 'shopping list' of priorities. Just as, clearly, other species may not – so far as we can discern – merit such priority treatment. For lack of adequate conservation resources, and *for*

*Coral reefs are among the most beautiful and fragile
of all natural wonders. Siltation, dynamiting for
harbours and exploitation to provide tourist souvenirs
are destroying them. Their myriad brightly coloured
fish are being trapped to meet demand from the huge
trade in tropical fish.*

no other reason, certain species will come pretty far down on a hierarchical ranking of priorities. Still others will be placed so far down on the list that they will effectively be consigned to a category that we designate 'We wish we could do something about them, but to our massive regret we just do not have the means available'.

This is *not* to say that we consign any species to a rag-bag collection of 'species that are not worth saving'. All species are worth saving: we cannot save them all. No species is without its intrinsic scientific interest. No species is without biological value. No species does not make a contribution of some sort to its ecosystem. We cannot possibly tell which species may offer economic potential to society at some stage in the indefinite future. Any species may one day generate aesthetic appeal in ways that we do not yet suspect. Still further justifications may be advanced in favour of any species – any species at all. And as a final argument, we can say that no species needs any justification for its survival, in so far as all species, being manifestations of life's diversity on Earth, can be considered to possess, *ipso facto*, a 'right to live'.

Norman Myers

Chapter 6

What's Happening to our Water?

108

The Julierpass, Switzerland. Mountain streams like this one are an essential link in the global cycling of water that keeps us constantly re-supplied with clean water for domestic or industrial use.

For centuries mankind has used the water available to him in rivers, lakes and streams. It has provided both food and drink, and been used for recreation, transport, energy, cooling, waste disposal and much more besides. Man-made lakes and reservoirs have been created, rivers dammed, watercourses altered, canals and aqueducts constructed.

Inevitably, the quantities of water available have affected what has been done with it. And, equally inevitably, human activities have had their effects upon the waters. Luckily, the hydrological cycle – in which rivers flow into the sea, and are continually renewed with clean rainwater – cleans up surface water naturally. And if the water has become so contaminated that the natural cycle cannot deal with the pollution quickly and effectively, then scientific purification techniques can be used; they are cheap, efficient and effective. Effective, that is, at the moment.

There have, of course, been times when things did not work as well as this. During the reign of King George III, a British member of parliament wrote a letter to the Prime Minister complaining about the appearance and smell of the River Thames. The letter, it is said, was written not in ink but in water taken from the river.

In the Thames, things have improved. Fish have now returned to central London, where for decades there was little if any life in the river. But a success story in London may not be so easy to repeat elsewhere; nor, for that matter, is it certain that a clean Thames can be maintained in the face of a growing band of industrial polluters. The fate of the Rhine in West Germany does not bode well for the future. Salmon and trout used to be caught there but have now disappeared, along with seven other migratory species of fish.

As the years go by, humans are making increasing demands on the world's water. Industry, agriculture, irrigation, mining, power generation and the concentration of millions of people into cities watered by one or at most two rivers, all contribute to a level of pollution that threatens to outstrip human ability to clean up afterwards. One of the greatest threats is the spread of water-borne diseases, especially in tropical countries.

All fresh water contains dissolved materials such as phosphates, gases such as oxygen, organic compounds, suspended particulate material such as silt, and micro-organisms. The quantities of each vary greatly from one area to another. But a lack of balance between them, or a dramatic increase in any one of them, can lead to aquatic chaos in which the whole ecology of the water body is upset. Then the water becomes unfit for human consumption, and some – or all – forms of aquatic life are killed. Both effects are becoming increasingly common.

Studies of water and how best to manage it have recently given birth to

Blackstone River, Massachusetts. Gross pollution is often the result of man's use of water. Each year about 3,000 square kilometers of water is withdrawn for human use. This is roughly equivalent to a lake the size of Europe as far East as Moscow and deep enough to reach half way to the Earth's core.

more efficient techniques for treating polluted water and cleaning up some of the worst contamination. At the same time new research has revealed a range of pollutants in rivers which had not previously been suspected. One result has been that nations are beginning to take the state of their fresh water more seriously than before.

By mid-1982, 323 water-monitoring stations in 50 countries had begun to report regularly on the pollution levels they were able to detect in fresh water. They were reporting data on more than 50 chemical and physical properties of water taken from rivers (monitored by 203 stations), lakes (monitored by 49 stations) and aquifers (monitored by 71 stations). The data were collected by a global monitoring centre in Canada. Its findings are just one indication of the increasing effect that human activity is having on the Earth's great natural cycles – in this case, the hydrological cycle in which the Sun's warmth evaporates water from the seas and the land, and releases it later as rain and snow in different places.

The Earth's Water Balance

Some of the water stored as ice in the middle of the Antarctic may have been frozen there for as long as 200,000 years. But most of the Earth's water participates in a vast and endless cycle that provides all the world's fresh water, and acts as a giant purification system. This system depends on just one simple fact: when water evaporates, the pollutants it contains do not go with it. This is because nearly all pollutants evaporate at much higher temperatures than water. As a result, evaporated water is pure. But the pollutants in what is left behind become increasingly concentrated.

Altogether, there are about 1,400 million cubic kilometers (km³) of water on the Earth. Some 97 per cent of it is sea water, nearly all of it permanently held in the oceans. The rest is fresh water, and at any one time 77 per cent of it is stored in the ice caps and in glaciers; a little over 22

per cent is ground water stored beneath the surface of the Earth; and 0.035 per cent is held in the atmosphere. Surprisingly enough, visible surface water – such as rivers and lakes – accounts for only 0.33 per cent of all the fresh water on the Earth.

The hydrological cycle starts with evaporation. Every year, 445,000 km³ of sea water and 71,000 km³ of surface water from the land are evaporated by the Sun's warmth. When this eventually falls again as rain or snow, it falls on the land and sea in different proportions: now the continents receive 104,000 km³ of water, and only 412,000 km³ return to the oceans. The extra 33,000 km³ which fall on the land eventually run back into the sea – but, in the meantime, those who live on the land where it falls have the benefit of it. Its distribution over the Earth's surface, however, is by no means even.

With the possible exception of just two countries, every country in the world does actually receive enough rainfall to provide for the basic needs of its population. But, when the supply is small and inconsistent, catching it and storing it are disproportionately expensive. In arid areas there may be no rain at all in some years. The humid tropics, by contrast, are deluged with water regularly every day during a precisely-defined rainy season; in some places, the onset of the day's rain can be reliably predicted to within a few minutes. In fact, the distribution of rainfall is so uneven over the Earth's surface that one third of all surface water is carried into the sea by just fifteen rivers; the Amazon alone carries 15 per cent of it.

Where the water actually falls is vitally important. Average world rainfall is 875 mm a year. Just over one quarter of this falls on land. But whereas rainfall in the United Kingdom averages 1,000 mm a year, for instance, Mauritania received just 2.5 mm in 1982. When there is as little water as this, it is difficult for humans and animals to find enough to drink. Water for washing and sanitation becomes a luxury. Industry cannot exist without at least

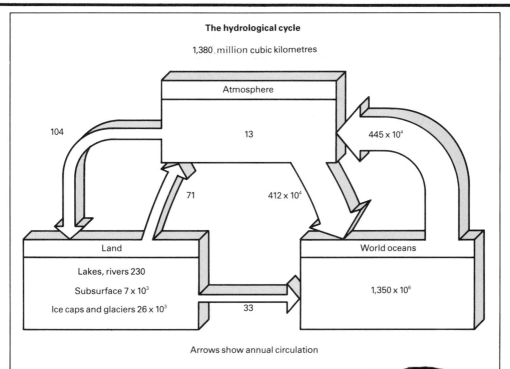

The hydrological cycle

1,380 million cubic kilometres

Atmosphere

104

13

445×10^4

71

412×10^4

Land

Lakes, rivers 230

Subsurface 7×10^3

Ice caps and glaciers 26×10^3

33

World oceans

$1,350 \times 10^6$

Arrows show annual circulation

some water and irrigation, of course, is out of the question.

In 1980 the amount of water used by mankind, worldwide, was estimated to be between 2,600 and 3,000 km³. Irrigation accounted for much the largest part (73 per cent), industry used 21 per cent, and domestic and recreational uses amounted to only six per cent. By 1985, it is expected, water use will have risen to 3,750 km³ a year.

Nor Any Drop to Drink . . .

In India, three children under five die from diarrhoea every minute. The disease is usually contracted from drinking polluted water.

India's rural population numbers more than 500 million people. Of these, 98 per cent have no form of sanitation, and almost 70 per cent have no safe supply of drinking water. Where water must be fetched from miles away, or bought from a carrier, it is far too precious to squander on washing and is kept for

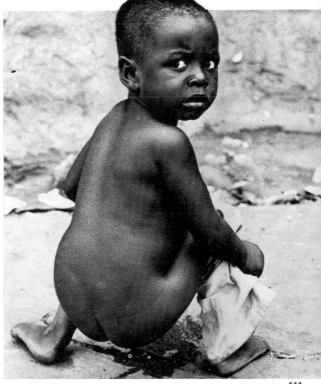

Top: Over 99 per cent of all the water on earth is locked away in the oceans and the ice-caps. Without the less than one per cent available in the atmosphere, in rivers and lakes and in underground stores, human life would be impossible. The purity and availability of this water is maintained by the constant cycling of water through the hydrological cycle.

Above: Lagos. Seventy-five per cent of people in the developing countries do not have adequate sanitation. About seven million children under five die each year from diseases resulting from poor sanitation.

drinking and cooking.

India's villagers, sadly, are not atypical. More than three quarters of the diseases in the Third World are probably due to dubious water supplies or lack of sanitation. According to estimates made in 1980 by the United Nations Environment Programme (UNEP), only 29 per cent of those in rural areas – worldwide – have access to safe drinking-water. Even in the cities and towns of the world, only 75 per cent have safe supplies of drinking-water. Despite efforts to provide more people with water and sanitation, the numbers of people with no clean water to drink increased between 1975 and 1980 by 100 million, and those without adequate sanitation by 400 million.

'All peoples . . . have the right of access to drinking-water in quantities and of a quality equal to their basic need,' stated the U.N Water Conference in 1977. And there is no doubt that efforts have been made to improve the situation. The current decade, 1981–90, has been launched as the International Drinking-Water Supply and Sanitation Decade, with the aim of providing everyone with safe

PER CAPITA WATER AVAILABILITY (1971)

thousands of
cubic meters of
water per person

High (more than 10)

Medium (10-5)

Low (5-1)

Very low (1 or less)

112

Where water actually falls is vitally important.
Global water demand currently stands at about 3000
billion tonnes a year, or about 700 tonnes per person.
We need water for domestic use, for industry and,
above all (about 73 per cent), for agriculture.
Insufficient water prevents economic development.

water supplies by 1990. The technologies to achieve this already exist and are relatively cost-effective and simple.

Yet by the time the Decade had begun, hopes that the aim would be achieved were already fading. Estimates of the cost of providing clean water for everyone were put at between 300,000 million and 600,000 million dollars over the decade – which would mean spending money about 100 times as fast as is currently the case. For the most part, Third World countries have responded enthusiastically to the theme of the

Decade. In India, for example, the government plans to supply water by the year 1990 to the 190,000 rural villages that have been identified as being most in need. However, this campaign will provide only a quarter of the rural population with the water it needs.

Even when money, skill and all the goodwill in the world are available, simply providing a village with water taps and latrines may not be enough. If poverty is such that simple survival is a struggle, the social organization needed to keep a sophisticated water supply system working is unlikely to exist. It is no accident that developed countries are usually those with abundant water supplies. Experience has shown that water supply schemes are more likely to work when they are integrated with other community development schemes. Yet this is still rarely the case.

Simon Watt, a British hydrologist, has described how water systems can break down, using as an example a system installed in Ban Ut Kwang, a village in north-east Thailand, in 1970.

With the assistance of US aid, a system was installed to pump treated water from a river one kilometer away from the village to a header tank in Ban Ut Kwang, and a network of supply and stand pipes was constructed. The system cost

Simple, cheap and reliable pumps, like these Mark II handpumps in rural India, are the most urgent priority for bringing reliable supplies of safe water to the rural poor.

80,000 dollars and was welcomed by the villagers; Ban Ut Kwang was a go-ahead modern village with a new school and its own health clinic, and was soon to have electricity.

Nevertheless, by 1977 the water supply system had been abandoned. Each of the 405 village families had been paying the equivalent of about $0.4 a month for their water supplies, but this figure did not include provision for the cost of maintenance. When the pump broke down there was not enough money to mend it. Arguments dragged on in the village about who should pay more in water rates – those with larger families or those with larger incomes. Meanwhile, the water system stood idle. Those who could afford it then began to buy water from a carrier. The poorer families went back to fetching their water from the polluted river.

Many international development experts now believe that success in supplying clean water to rural communities depends on the design and availability of a simple, cheap and reliable hand pump to bring ground water to the surface. Too many villages, like Ban Ut Kwang, have rusting abandoned pumps awaiting repair. The aim is to design a pump that could be maintained – by a villager appointed to take care of it – using just one spanner. New designs and more stringent tests, together with the use of new plastic components, may bring success nearer.

But if rural villages are to depend on pumping ground water to the surface, more attention will need to be paid to keeping underground sources free of pollution. The main attraction of using ground water rather than surface water is that it is much less likely to need treatment before it can be drunk. However, prospects for the world's stores of underground water are not bright.

Protecting Underground Supplies
The water-table lies at the upper level of the zone of ground water below the

Earth's surface. Where there are lakes and rivers, the water-table is near the surface. But although ground water is usually clean and pure, it is by no means sealed off from human activity – and nor is it necessarily static. Ground and surface water are linked together by a series of continual interactions – interactions that serve to keep underground supplies topped up but that may also be a source of pollution.

The rate at which ground water is renewed depends on many factors – but the nearer it is to the surface, the faster it is renewed. Near the water-table itself, renewal may take as little as a year; in deep underground aquifers, renewal may take thousands of years. If these deep water stores – which are virtually underground lakes of 'fossil water' – ever become polluted, their contamination is as near permanent and irreversible as makes no difference.

When ground water is over-exploited by too many wells, the water-table may be lowered as a result. Even if there is sufficient water to recharge the aquifer, it

Top: acid rain (see page 100) is a growing problem in some countries. Acidic water can dissolve heavy metals from soil and water pipes, thus contaminating drinking-water supplies. It can be controlled, by liming lakes, here being carried out from the air.

may be only party replenished if it is made of clay or soil with variable porosity. In this case, once it dries out, it will shrink and be unable to hold as much water as it did previously.

In this situation, precious water is not the only thing to be lost. If underground clays and other materials shrink sufficiently, there is likely to be land subsidence at the surface. This can have serious consequences. Better planning, and improved understanding of the links between surface and ground water, are needed to avoid such catastrophes.

The Salinity Factor

Previous chapters have already mentioned the problems of excessively salty soil caused by inefficient methods of irrigation. What happens is that the rain that filters down through the layers of damp soil, sand and clay to recharge underground aquifers, dissolves and collects different salts on its way. Most ground water therefore contains some salts. These salts may be brought to the surface through the irrigation of badly

Above: a forest in Colombia, killed by saline water.

drained land. The water-table rises and eventually the land becomes waterlogged. Salts from the ground water can then rise to the surface, where evaporation concentrates them. In the end, saline or alkaline soils will be formed on which little will grow and yields will continue to drop until the land is either abandoned or effectively drained.

The same thing can happen when large areas of land are cleared of forest. The thin soil is then directly exposed to rain, which consequently penetrates much deeper into the ground. Eventually, the land can become waterlogged and saline.

Soil salinity is a serious threat to agriculture, and it occurs all over the world in irrigated lands. By the mid-1970s an estimated 952 million hectares (about 3.7 million square miles) of the world's soil was affected by salt, but this includes badlands and deserts. Perhaps one tenth of all irrigated land is waterlogged, saline or both – 22 million hectares (around 85,000 square miles) in total.

If ground water in coastal areas is over-exploited and the water-table lowered, the water may become contaminated by sea water. This is a severe problem in arid areas but salt-water contamination is by no means confined to them: it was recorded in the United Kingdom in the mid-19th century when it was caused by overenthusiastic extraction of water in both London and Liverpool. The same thing has since happened in Japan, the United States and a number of European countries.

Where salinity infiltrates ground water near the coasts, attempts have been made to minimize the effects by what amounts to constructing a 'barrier' of fresh water. This can be done, for example, by diverting rivers over a wider area to encourage infiltration, by excavating basins and filling them with water, by building dams, and by digging ditches and furrows. These techniques often work and also provide a useful means of storing excess water when rainfall is plentiful for use in the future for irrigation.

The Threat of Pollution

Pollution from radioactive wastes, toxic materials and heavy metals cannot be so easily prevented. Some aquifers, which are relatively safe from surface pollution because they are covered by a layer of impervious rock, may nevertheless become polluted if it is decided to bury dangerous wastes deep in the ground, or to use old wells as a dumping ground for such waste.

Accidental pollution from oil or other hydrocarbon fuels occurs when pipelines leak or break. Oil, once in the soil, is persistent and difficult to remove. However, even these accidental spillages are not as serious as the threat from the use of artificial fertilizers.

Phosphates from fertilizers are not usually detectable in ground water because they mostly become absorbed in the soil moisture or decomposed before the water carrying them through the soil reaches the water-table. Nitrates, however, are found in ground water in increasing quantities and in many rural areas the water must now be treated before it can be regarded as fit for human consumption.

Pesticides and herbicides are also finding their way into ground water. One

Signs like these have become more frequent as pollution has outstripped both nature's ability and man's willingness to keep water clean.

of the problems with this kind of pollution is that is may accumulate in considerable quantities in the soil before its presence can be detected in the water it finally pollutes. By then it is too late. Many pollutants may be percolating their way through the soil to deep aquifers while the water itself still appears to be free of contamination. As a result, more importance is likely to be attached in future to regular monitoring of soil moisture for pollutants.

'Acid rain', too, is now widely recognized as a threat to ground water supplies. This rain contains nitric or sulphuric acids formed from the oxides of sulphur and nitrogen which are emitted by power stations, industrial chimneys and vehicle exhausts. These oxides may be transported 1,000 kilometers or more in the atmosphere before finally falling to Earth in rain or snow. Some domestic water supplies in Scandinavia, where acid rain pollution is now very severe, have been found to be very acidic. One of the problems with this type of pollution is that it is essentially unpredictable – but recent international legislation may eventually reduce the incidence of acid rain in Europe by controlling sulphur and nitrogen emissions from power stations and factories.

The increasing pollution of ground water has led to suggestions that drinking-water supplies should be more intensively treated in the future to eliminate pollutants that may be dangerous to health. For the time being, however, as water sources go, ground water is still relatively clean. The immediate problems come from drinking-water taken from open surfaces such as lakes, reservoirs and rivers.

The Lakes and the Rivers

Colombia's Bogota river (often claimed to be one of the world's most polluted rivers) collects sewage and industrial effluents for almost the entire length of its long course towards the sea. Analysis of its water – which villagers downstream have

no choice but to drink – reveals a very special cocktail containing mercury, copper and arsenic. A recent survey of fifteen Mexican rivers has found them unfit for most purposes.

Mercury and asbestos are increasingly common constituents of river water, and there are suspicions that they are already causing human disease, including cancer. Chlorination, too, has been blamed for producing carcinogenic compounds in domestic water. The problem is such that economic experts have calculated that the reduction in ill health that might be achieved by a much more intensive filtration of domestic water supplies, using carbon filters, could well justify the cost.

However, costs are involved other than those of human health. Recently, attempts have been made in the United States to put a price on the benefits to mankind of keeping domestic water cleaner. Improving the quality of drinking-water in the United States, it is estimated,

A sewage treatment works. Proper sewage treatment can remove most of the organic material from used water. But nitrates and phosphates remain in the water and cause new problems.

could save the country as much as 2,000 million dollars a year. Treating water used by industry could save another 400 to 800 million dollars a year. And if the water used by commercial fisheries could also be improved in quality the benefits might reach 1,200 million dollars a year. Crop yields, too, are affected by the quality of the irrigation water used – but so far no one has managed to calculate the benefits

untreated sewage is discharged into a lake, micro-organisms in the water get to work to break down the organic matter it contains. But to do this they need oxygen. And the oxygen they have to use is the small amount that is always dissolved in water, and on which fish and other aquatic animals depend for their survival. Of course, this oxygen is quickly renewed in shallow or rough water. In water that is

of improved supplies because the other factors involved, such as the changing weather, have a much larger effect.

The story of Lake Erie on the United States/Canadian border illustrates just how complex the chain of contamination in fresh water ecology can be. By the mid-1960s pollution there had become so severe that nearly all the popular beaches had been abandoned, and enormous piles of decaying fish and algae piled up on the shores of the lake every summer. One particularly attractive inhabitant, the mayfly, had disappeared altogether. The reasons for all this turned out to be much more complicated than at first appeared.

One critical factor in the health of water relates to what is called 'biological oxygen demand' or BOD. If, for example,

deep and still, however, new oxygen is harder to come by, and oxygen levels drop steadily while the organic material is being decomposed on the lake bottom. This is why fish life disappears so quickly from water in which sewage and other untreated organic matter is dumped.

However, while untreated sewage probably does enter Lake Erie, most human waste passes through sewage plants first. Their job is to decompose the organic matter before it enters the river or lake. One stage in the treatment involves getting as much oxygen as possible to dissolve in the water, which is done by gently spraying the water over the sewage beds. What finally emerges from a sewage plant should be clean, clear water from which all organic material has been

Eutrophication, caused by a build up of nutrients in the water, is an all too visible and unpleasant consequence of gross pollution.

removed and converted to nitrates and phosphates.

Research on Lake Erie, however, soon showed that these inorganic salts are not as harmless as was thought. Ultimately, they are responsible for building up a huge volume of organic material in the lake. The end effect is almost the same as dumping untreated sewage in the lake. What had been happening in Lake Erie was this.

Every summer, as the lake warmed up, huge algal 'blooms' appeared on the lake, turning parts of it into something the colour and consistency of pea soup. These blooms were caused by the presence of all the artificial fertilizer in the form of phosphates and nitrates released into the lake from surrounding sewage plants. But the algae died as quickly as they grew. Once dead, they sank to the bottom, forming a thick layer of organic matter on which micro-organisms then began to work. In doing so, they depleted the oxygen levels of the lower layers of the lake. Fish, seeking the cooler waters of the lake bottom, began to die from lack of oxygen. And the mayfly nymph, which lives on the lake bottom, also died.

This process is called eutrophication, and by the 1970s had become a common occurrence in inland waters. However, as scientists have learned more about it, attempts to control it have become more successful. In some cases, as in Lake Washington in the United States, all that was required was to divert the sewage input elsewhere, after which the lake promptly recovered. However, when massive quantities of nutrients accumulate in the sediments, clearing up the problem is more difficult. The sediments may have to be dredged – a solution that is only practicable in small lakes.

Another problem is increased nitrate levels created by run-off into rivers from nitrogen fertilizers. It is not recommended for drinking-water to contain more than 10 to 12 mg of nitrate per litre – yet concentrations of this order can now be found, for instance, in rivers in the United

Kingdom. Twenty years ago this was not so. Nitrate is not believed, in itself, to be harmful to human beings. But it can be converted to nitrite by the action of intestinal bacteria, especially in young children. Nitrite is poisonous, and prevents the transportation of oxygen in the blood; infants affected turn blue and are in serious danger of death by asphyxiation.

Although it is not easy to regulate the way in which chemicals from fertilizers and pesticides enter waterways, controls can be applied to industry. In some countries deteriorating water quality has been checked by insisting that waste industrial water is treated before it is discharged. However, advances in detecting pollutants in water have shown that, even after treatment, some industrial

Helicopter spraying fertilizers on the Mziuri state farm in the Galsky region of the Soviet Union. The run-off of nitrates from fertilizers is a growing problem in many parts of the developed world.

discharges result in small concentrations of chemical pollutants. Such discharges may be modified, but not removed, by the treatment they later receive in a municipal water supply system.

Pollution from uncontrollable sources has lately decreased in developed countries through treatment of wastes before discharge; but the less controllable areas of pesticide and fertilizer pollution are still causing concern, and pollution from them is increasing.

The Oceans

In the 1960s, 41 people died and 70 more were seriously ill after eating fish caught in Japan's Minimata Bay. The sickness (which came to be known as 'Minimata disease') was caused by concentrated mercury in the fish – up to 15 mg per kilogram – and the episode, together with other similar problems all over the world, alerted public opinion to the dangers of polluting the oceans.

Mercury occurs naturally in sea water (its presence is due to the erosion of mercury-containing rocks) but in almost insignificant quantities. Where rivers carrying waste waters from the plastics, paint, chemical and paper industries flow into coastal waters, however, levels may be far above the normal – especially if the seas are enclosed, or semi-enclosed, by land. Fish then concentrate the mercury: tuna and swordfish have been found with a hundred times as much mercury in their bodies as the surrounding water contains. In 1980 Mediterranean tuna contained around 1.26 mg per kilogram of mercury – more than four times as much as is found in fish caught in the Pacific Ocean.

Nowadays, levels of metal pollution in mussels in coastal waters off the United States are measured regularly. Since the mussels concentrate such pollutants in their tissues, they make a useful barometer of the health of the coastal waters. The presence of lead, cadmium, zinc, copper, nickel, plutonium, caesium, halogenated hydrocarbons and petroleum hydrocarbons can be regularly checked for in this way. If any of them are present, the mussels will find it. This is one reason why eating shellfish from polluted waters is potentially so dangerous.

DDT, a pesticide which has been in use since 1942, can – like mercury – be concentrated in the bodies of birds and fish. It has been found in sea water, rainwater and Antarctic snow, far away from areas where it is used. Levels of DDT

Minimata, a bay in Japan, has become a watchword for the tragic consequences of pollution.

in the sea in the northern hemisphere once caused concern but have fallen over the past ten years; in the southern hemisphere, however, where DDT has continued to be used, the amount in the oceans may have risen, and some marine organisms contain potentially dangerous levels.

Everything that is carried away by rivers ends up in the sea. This means that

that pollution has to be blamed for the decline in numbers of certain species of aquatic mammals.

The attempt to do something about this must rank as the most successful environmental campaign of the 1970s. That the state of some 'inland' seas has improved is due to the initiative of the United Nations Environment Programme which set up its Regional Seas Programme

coastal waters are invariably polluted by sewage, agricultural chemicals and oil as well as by heavy metals and pesticides. The most obvious effects can be observed in the semi-enclosed seas such as the Gulf of Mexico, the Mediterranean and the Baltic Sea. In places, contamination is so severe that the sale of fish has had to be banned. In others, experts are now sure

in the mid-1970s. The first area to be tackled was the Mediterranean. There is no tide in the Mediterranean, and its waters are renewed only every 80–100 years. It is polluted with wastes from industry, sewage, ships and agriculture – and matters are made worse by its high tourist population. The condition of its waters has been blamed for local

The sea is the final dumping ground for much of our waste and pollution.

outbreaks of hepatitis, dysentery, meningitis and cholera.

The major success of the Programme was in persuading the sixteen countries involved – some of which were mutually hostile nations – to co-operate in cleaning up their common sea. When the Mediterranean Action Plan ('the Blue Plan') was signed by all the relevant nations in 1975, the Programme set in train negotiations to achieve similar plans in nine other polluted, semi-enclosed seas: the Red Sea, the Kuwait region, the West and Central African coastal seas, the wider Caribbean, the East Asian seas, the South-East Pacific seas, the South-West Pacific seas, the South-West Atlantic and the East African seas. By 1983, Plans of Action had been signed for most of these.

There is, however, a vast difference between coastal waters and enclosed seas, and the open ocean itself. While such famous figures as Jacques-Yves Cousteau and Thor Heyerdahl have often denounced the pollution of the oceans, the evidence suggests that contamination of the open ocean is still, fortunately, very limited. A 1982 report on marine pollution by experts from eight UN organizations concluded that the open oceans are healthier than might be expected:

In the open sea we have not detected significant effects on the ecosystem. Trends have indeed been observed of the concentrations of several contaminants, some up, some down, but these are not reflected in environmental deterioration. Nevertheless, general trends of increasing contamination have been recognized in some areas. These trends are warning signals, noticeable mainly in the coastal waters used most intensively by Man. The oceans are

It is often the places to which people most want to go — like this Mediterranean beach — that are the most polluted.

capable of absorbing limited and controlled quantities of wastes and represent an important resource. But . . . the effects of pollution should be carefully monitored.

Another recent observation has been of the quantities of radioactive material in the sea (strontium-90, caesium-137, plutonium-239 and plutonium-40) from nuclear-weapons testing and, to a smaller

1. Mediterranean Region
2. Kuwait Action Plan Region
3. Caribbean Region
4. West and Central African Region
5. East African Region
6. East Asian Region
7. Red Sea and Gulf of Aden Region
8. South-West Pacific Region
9. South-East Pacific Region
10. South-West Atlantic Region

The same report, while noting that in some inter-tidal and sub-tidal communities oil has penetrated sediments and recovery can be expected to take years or even decades, remarks that oil spills rarely have effects as drastic as used to be feared. Birds suffer severely but oil does not seem to pose a threat to the survival of species, to have lasting effects on coastal ecology or to cause anything other than temporary disruption to coastal economies.

Mankind's understanding of inter-changes between the oceans and the atmosphere has recently advanced; as well as the concept that the oceans can absorb excess carbon dioxide produced by Man's activities (such as burning fossil fuels and deforestation), it is now believed that the oceans may also absorb, via the atmosphere, other contaminants such as metals and synthetic chemicals.

extent, from nuclear power production. Radionuclides of artificial origin in the oceans total about 0.1 per cent, but concentrations are much higher around points of discharge. Sellasield (formerly called Windscale) has thus left the Irish Sea with significantly more radioactive material than it once contained, and some of the material discharged from Sellasield has been traced in a journey around the north coast of Scotland, across the North Sea and around the Norwegian Sea.

Traditional open access to the sea and its resources has already ended. More than a hundred countries already claim authority over the area of sea up to 200 nautical miles from their coasts. This, it is believed, will afford opportunities for more rational and efficient management of the fisheries – recognized, these days, as a major food resource.

The coastal zones of the oceans are amongst the most biologically productive ecosystems. They are also the receptacle for huge amounts of polluted water from the land and the site of important industrial activities such as oil and gravel extraction. In these zones, and especially in the enclosed or semi-enclosed seas, pollution has risen alarmingly. In the mid-seventies the United Nations Environment Programme established its Regional Seas Programme to begin tackling these problems.

World Fisheries

Between 1900 and 1962, the total annual fish catch from both marine and fresh waters rose by a factor of eight, to about 40 million tonnes. In 1976, more than 70 million tonnes of fish were landed throughout the world. Thirty-two countries depend on sea food for one third or more of their protein intake, and fish provides 18 per cent of all the animal protein consumed worldwide. (It provides 6 per cent of the world total protein intake.)

The quantities of fish taken from the sea – as opposed to inland waters – seem to have levelled off in the past ten years at around 60–65 million tonnes. Many people now believe that fish stocks have been over-exploited, and the presence of smaller quantities of some of the 'high quality' species in the catch would seem to support this view. In 1970, herring, sardine, anchovy, cod and haddock accounted for about 32 per cent of the catch; by 1975 their proportion was down to 25 per cent.

Peruvian anchovy catches fell from about 12 million tonnes in 1970 to only 4 million tonnes in 1972 and have failed to recover. And the herring catch in the North Atlantic, after increasing until the mid-1970s, has lately decreased severely. The causes are taken to be over-exploitation, plus an element of natural change. The Peruvian fishery is known to be particularly sensitive to variations in current and nutrient patterns; and the herring catch in northern Europe is known

The dumping of radioactive waste at sea has aroused fierce protest from environmental organizations.

to have fluctuated widely ever since the Middle Ages. For the catch to be maximized without over-exploiting marine resources, it is obviously important that interactions such as these can be understood. Some believe that, with better management, the world catch could sustainably have been up to 20 million tonnes a year higher than it was throughout the 1970s.

With encouragement, fish stocks may be built up once again – and if fishing is controlled, the production of readily marketable fish may be as high as 90 million tonnes by the turn of the century. Because this will not be enough, it is expected that the catch will come to include previously unexploited species – such as krill.

extracted from the Southern Ocean. Although this figure is now believed to be far too high, experimental fishing for krill is under way, in an attempt to determine how much might be yielded. Uncontrolled extraction is unlikely to be permitted; krill is the main food for five species of great whale – including endangered species – and for some seals, seabirds and fish. Already a convention is being drawn up to regulate removal of living resources from the Southern Ocean.

But, says *The World Environment 1972–1982*, a report by the United Nations Environment Programme, 'the extension of economic zones to 200 nautical miles will not provide adequate conditions for protection or rational exploitation of these resources unless combined with inter-

WORLD FISH CATCH, FOOD AND NONFOOD USES, 1950-79	Year	Food	Fish Meal	Total
			(million metric tons)	
	1950	18.1	3.0	21.1
Source: Food and Agriculture Organization	1955	24.3	4.6	28.9
	1960	31.4	8.6	40.0
	1965	37.2	16.3	53.5
	1970	42.6	25.5	68.1
	1971	43.9	24.6	68.5
	1972	45.0	19.2	64.2
	1973	47.5	17.6	65.1
	1974	48.9	20.6	69.5
	1975	49.0	20.3	69.3
	1976	50.7	22.1	72.8
	1977	52.8	19.7	72.5
	1978	52.8	21.0	73.8
	1979	52.5	20.9	73.4

Much has been claimed for krill (a small shrimp-like creature found in quantity in the Antarctic); scientific estimates in the early 1970s suggested that up to 100 million tonnes of krill might sustainably be

national agreements and regulations. Proper assessment of the potential yield is a vital prerequisite to successful management, setting a balance between stock recruitment and catch potential.' For

Fish account for about 18 per cent of the animal protein in the human diet. But this average figure hides the fact that over 30 countries get more than a third of their animal protein from fish.
Over-exploitation has led to an end to growth in the

amount of fish caught. At least 25 of the world's most valuable fisheries are depleted. It is estimated that fish catches are 25 per cent smaller than they would have been if properly exploited.

efficient management, it goes on, agreements to regulate the catch of various species must be made, and must then be enforced.

Coastal wetlands, flood plains, seagrass beds, coral reefs and mangrove swamps will also need special protection. These are nursery and breeding grounds for many species of fish, but they are damaged by industrial and agricultural pollution, dams, silt and dredging. Mangrove swamps are disappearing, too, as they are cut for fuel; in India, since 1900, around 16,000 square kilometers (more than 6,000 square miles) have been lost. Coral reefs in Hawaii were damaged by blanketing deposits of sediment on them through coastal urbanization. The increasing tendency to build nuclear power stations and oil refineries, pulp mills and chemical works near shorelines to take advantage of the nearby water for cooling is in conflict with the needs of the marine population. Careful management can restrict to a minimum the disturbances to these important habitats, but wise planning is essential.

Have We Saved the Whales?

Many marine mammals and their habitats are also in need of protection. Historically, hunting was the major threat to mammals such as whales, porpoises, dolphins, seals, sea lions, sirenians, sea otters and polar bears. Now, as the technology to allow exploration of mineral extraction from the ocean floor is being developed, and mankind searches for more and more species to eat, they risk losing their living space and their food.

Whales, of course, have been severely endangered by hunting. It has been estimated that stocks of the ten species of large whales (including the blue whale, believed to be the largest creature ever to have lived on the Earth) which have formed the basis of commercial whaling in this century have been reduced to 48 per cent of their original populations. But this average figure is misleading. The reduction of some species has made

life easier for others and, despite hunting, they have proliferated. Other species have been reduced to 5 per cent or less of their former numbers.

The International Whaling Commission – comprising most whaling nations and other interested parties – was established in 1946 in an attempt to impose responsible management. It sets quotas each year for allowable catches of each species. Whale populations nevertheless continued to decline during the 1960s.

In the late 1960s, however, scientists publicly voiced their concern about whaling and the campaign was taken up by the general public (culminating eventually in the dramatic ramming and blowing up of the pirate whaling ship *Sierra* in Portugal in 1980). In 1970 the United States banned whaling and the import of all whale products. This open support gave new strength to the International Whaling Commission and hunting quotas were rapidly reduced from

Krill, the Antarctic shrimp on which the great whales feed, may also become food for humans. Some 15 nations are engaged in research on, or experimental fishing for krill, taking about 150,000 tonnes a year from a potential that could be as high as 100 million tonnes.

46,000 whales in 1973 to just 14,000 in 1982. Although the total ban on whaling desired by many has never come to pass, the quota for the severely depleted whale was reduced to zero for the 1982 catch, despite an objection from Japan.

It is believed that action was taken in time, and that stocks of the endangered whales will gradually recover – although they may never again exist in their original numbers. Most, however, will probably recover suffiently to allow them to continue to be hunted – despite the distaste that many people have come to feel for the killing of these intelligent creatures.

Death of a sperm whale.

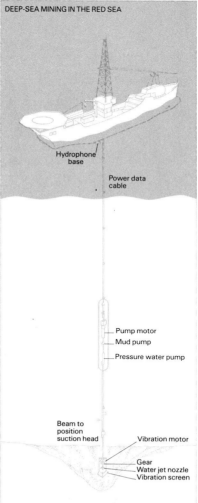

DEEP-SEA MINING IN THE RED SEA

Hydrophone base

Power data cable

Pump motor
Mud pump
Pressure water pump

Beam to position suction head

Vibration motor

Gear
Water jet nozzle
Vibration screen

Top: the Beryl A production platform in the North Sea. The endless quest for oil, driving exploration into ever deeper and more dangerous seas, poses one of the most serious pollution threats to the ocean.

Above: the sediments of the Red Sea, two kilometers below the surface, are rich in minerals including zinc, silver and copper. The mud is broken up with a vibrating screen and jets of seawater, then pumped to the surface where the metals are extracted. In other parts of the ocean, nodules rich in manganese and nickel may be extracted from the sea bed.

Our Last Unhurt Resource?

By a miracle, centuries of pouring waste materials into the oceans do not seem to have done serious damage. Perhaps mankind's increasing environmental awareness will permit the seas to escape the fate that has befallen some of the lands of the Earth.

In fact, the seas, the rivers and the lakes are connected together by a curious kind of web. The inland waters have been grossly polluted at times when we failed to understand what was happening to them. Things have now improved in the developed world, although they show no signs of doing so elsewhere. But inland pollution is, in a sense, a trivial matter – after all, we are using the rivers mainly as a means of transport, as pollution canals, to take our waste out to sea. If they occasionally become overburdened with rubbish in the process, matters can always be rectified.

But once the rubbish arrives in the ocean, the situation changes. We cannot pour the oceans out into a yet larger body of water to clean them up. What goes into the Pacific and the Atlantic, the Caribbean and the China Sea, stays there. For good.

So far, it seems, we have not overstretched what has been called 'the last resource'. But we have done damage. As the Chief Scientist to the UK Ministry of the Environment told an environmental conference in June 1982: 'We have a patient on our hands that can be saved and for whose health we should not yet despair.' An American oceanographer took the argument a stage further. 'With the oceans we have had an opportunity to be wise', she said, 'and we need not repeat errors committed on the land. But we must hurry.'

Robin Clarke

'There is but one ocean, though it has many coves.'
David Brower – founder of Friends of the Earth.

Chapter 7

Atmospheric Pollution

A good many excellent films have been set in the stinking yellow smog for which London was famous. Detectives, crooks and call-girls have all battled their coughing way through the swirling fumes, which in spite of the ingenuity of the film studios never appeared to be anything but a cheap imitation of what Londoners used to know as a 'real pea-souper' – a phrase used to conjure up the particularly invidious green-yellow colour of the London fog at its worst.

In just five days in 1952, one of London s worst ever smogs killed more than 4,000 people – a tragedy that remains unsurpassed as the worst air pollution disaster. London smogs were caused by a combination of fog and pollution, the latter being mainly sulphur dioxide released when sulphur-containing coal and oil are burnt either domestically or in fossil fuel power plants. After the 1952 tragedy, the London authorities began to take more and more serious account of their polluted city. The whole city – like many others in the United Kingdom – was declared a smoke-free zone, and no one was allowed to burn a coal fire unless a pre-processed, sulphur-free fuel or a more efficient 'smoke-eating' stove was used. Industrial chimneys were required to be higher, and to have built-in dust precipitators to remove some of the worst of the pollution.

The results were spectacular. The London smog is now a thing of the past; there have been no serious outbreaks of air pollution in London since the early 1960s. A problem that began around the year 1600, and had been building up ever since, was virtually cured in less than a decade at the beginning of the second half of the 20th century. Similar successes have attended most of the large cities in

Fleet Street, London, 1952, in the grip of the great smog which first drew the world's attention to the seriousness of air pollution problems.

the developed world.

The story in developing countries, however, is another matter. There, poor populations have had to continue to depend on coal or, more often, fuelwood or charcoal for their cooking and heating needs. Controls have either not been introduced, or have proved either impossible or inhumane to implement. Cities like Calcutta and Dacca are more polluted than they have ever been. One measure of the degree of pollution of the atmosphere is to count the number of what are called 'suspended particles' in the atmosphere. During 1973–1977 there were sixteen times more suspended particles in the air in Calcutta than in London.

'I Shot an Arrow in the Air – It Stuck There' – LA Graffito

The Los Angeles smog is a different thing altogether. It was first noticed in 1943 when residents began to detect a whitish or yellow-brown tint to the air which stung the eyes. Comparing their problems with what they knew of London, the local authorities blamed sulphur dioxide and took steps to reduce it. They succeeded – but the smog got worse. Research later revealed that the problem came primarily from vehicle exhausts that emitted nitrogen oxides. These, in turn, were combining with other hydrocarbons in the atmosphere, such as fumes from the fuel itself, in a reaction that was activated by sunlight. Hence the term photochemical smog.

So efforts were now made to reduce hydrocarbon emissions from oil wells and refineries. The efforts succeeded, and hydrocarbon emissions were reduced from 2,100 tonnes a day in 1940 to only 250 tonnes a day in 1957. But the smog continued to get worse.

Eye irritation was reported in Los Angeles on 187 days of the year in 1959, on 198 days in 1960, on 186 days in 1961 and on 212 days in 1962. And not only in Los Angeles. Photochemical smog has been found in many parts of the world, particularly where there is dense traffic and a warm, sunny climate as there is in Lima, Mexico City, Sydney, Melbourne and Tokyo. In all these places visibility is seriously reduced and sometimes as little as one tenth of the available sunlight actually penetrates the smog and reaches

the ground.

Finally, however, Los Angeles hit on the solution to the problem: antipollution devices on car and lorry exhausts. Once they were introduced – despite the expected defensive shrieks from the automobile industry – the problem began to abate. Between 1965 and 1974 the monthly maximum oxidant concentration in downtown Los Angeles was reduced from 0.27 parts per million (ppm) to 0.17 ppm. In San Francisco Bay it fell from 0.13 to 0.09 ppm over the same period.

Controls were also introduced in Tokyo where the problem had become so bad that warnings were issued when pollution reached a certain level. The number of alerts there fell from 330 in 1973, to 150 in 1978 and to 84 in 1979.

Smogs, of course, are only the most visible aspect of air pollution. In the 1950s and early 1960s there was great concern about the levels of radioactivity in the atmosphere all over the world. Radio-active calcium and strontium were getting into grass, into cows and into milk. The

Los Angeles, USA. Despite some improvements, many cities are still seriously affected by photochemical smog produced by the action of sunlight on vehicle exhaust gases. Even in Los
Angeles, where the problem was first detected and where vigorous action has been taken, there are fears that smog will harm the performance of athletes in the 1984 Olympics.

Partial Nuclear Test Ban Treaty, signed in 1963, forbade the testing of nuclear devices in the atmosphere and put an end, in a single blow, to the whole issue. Radioactivity in the atmosphere has since retreated to its natural level, and this form of pollution is now rarely mentioned. Only France, to its everlasting shame, continues to defy the nuclear test ban, and to explode nuclear weapons in the atmosphere.

But there are other toxic chemical pollutants in the atmosphere, of which lead is currently the most important. Lead is poisonous, but is used as an anti-knock agent in petrol (gasoline) and enters the atmosphere from car exhausts. Near toxic levels have been found in traffic police-men, and pigeons living in Philadelphia contain ten times as much lead as their more fortunate rural brethren.

However, the toxic effects of lead, which include a possible lowering of intelligence in young children whose lead levels (measured in the bloodstream) become too high, have been so widely

Not all air pollution is visible. Lead is a particularly dangerous air pollutant, and vigorous campaigns, such as the CLEAR campaign in England, have sprung up in many countries to press for the elimination of lead from petrol. A number of countries, including the USA, Australia and Japan, already insist on all vehicles eventually running on lead-free fuel.

publicized that governments are taking action. The United States was a pioneer in legislation to lower lead levels in petrol. The United Kingdom stalled for some time, but is now acting also, among the first European countries to do so. Other countries will undoubtedly follow suit in due course. While lead may be an issue now, it is not likely to be so for much longer, thanks in part to an exhaustive report on the subject prepared in the UK by the Royal Commission on Environmental Pollution in 1983. Among many other things, this report put paid to the myth that opting for lead-free petrol would cost governments, petrol buyers and car manufacturers vast amounts of money – a claim that had long been perpetuated by those with vested interests in maintaining the *status quo*.

These aspects of atmospheric pollution appear highly encouraging. But there is usually a dark cloud to every silver lining, and air pollution is no exception. While we have been intent on cleaning up the mess on our doorsteps, we have been simply pushing the problem onto other people's doorsteps. And in the process we have created issues for the future so enormous that London smogs seem positively reassuring in comparison.

When the Rain Turned to Vinegar

Back in 1926, the Inspector of Freshwater Fisheries in Norway made a note that the sudden death of large numbers of newly hatched salmon fry in Norwegian waters seemed to be linked to water acidity. He may have been the first to record this effect. Fifty-four years previously, in 1872, the British chemist Robert Angus Smith had first mentioned 'acid rain' in his book *Air and Rain: the beginnings of chemical climatology* – but Smith never dreamed, when he coined the term, that 'chemical climatology' would, in the 1970s, produce

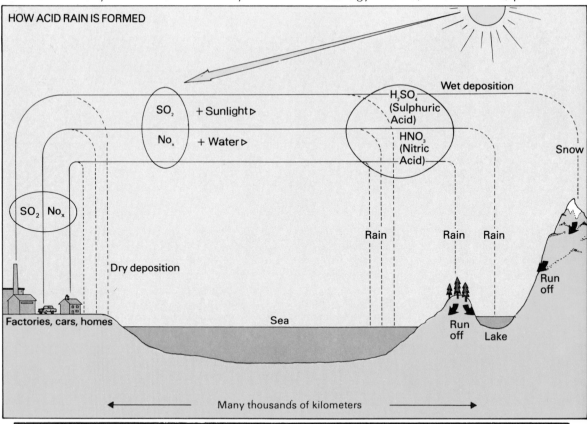

HOW ACID RAIN IS FORMED

SO$_2$ + Sunlight ▷

No$_x$ + Water ▷

H$_2$SO$_4$ (Sulphuric Acid)

HNO$_3$ (Nitric Acid)

Wet deposition

Snow

SO$_2$ | No$_x$

Dry deposition

Rain   Rain   Rain

Run off

Factories, cars, homes

Sea

Run off   Lake

Many thousands of kilometers

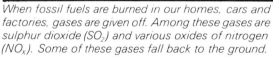

When fossil fuels are burned in our homes, cars and factories, gases are given off. Among these gases are sulphur dioxide (SO$_2$) and various oxides of nitrogen (NO$_x$). Some of these gases fall back to the ground. *they are* dry deposited. *Others interact with sunlight and the moisture in the air to form acids. These acids are washed out of the sky when it rains or snows as* wet deposition *or, more commonly, 'acid rain'.*

rain as acid as vinegar – rain which is now recognized to have destroyed all life in tens of thousands of lakes and is suspected of threatening the survival of vast areas of forest in both Europe and North America.

Acid rain is caused mainly by the emission of sulphur and nitrogen oxides when fossil fuels are burnt in power plants. The smelting industries and motor vehicle exhausts also produce these oxides. If they are swept up into the atmosphere by tall chimneys – some of which may be 400 metres (1300 feet) high – they can travel thousands of kilometers before returning to the Earth. The longer they stay in the atmosphere the more likely they are to be oxidized into sulphuric and nitric acids, which then dissolve in the water in the atmosphere and fall to the ground as acid rain (or snow).

The result is increased acidity in the soil, damage to crops and forests and frequently the death of all life in lakes. The effects have been particularly pronounced in Scandinavia and central Europe, and in Canada. Sulphur concentrations in rain doubled in central Europe between the 1950s and the 1970s and today each hectare of forest and farmland there receives an annual deposit of 30–60 kilograms (65–130 pounds) of sulphur and 15–30 kilograms (33–65 pounds) of nitrogen.

Damage from acid rain used to be confined to areas fairly near the source of pollution. But between 1960 and 1980 – a period during which the production of sulphur dioxide doubled in Europe – many power stations acquired chimneys that were more than 180 meters (600 feet) high. These reduced local pollution by pushing the stack gases up to higher levels in the atmosphere. But the gases were transported much further than before.

The 600ft. high chimneys of the British Central Electricity Generating Board's power stations have certainly helped to control local air pollution. But there is growing evidence that they may just be spreading the problem to other countries.

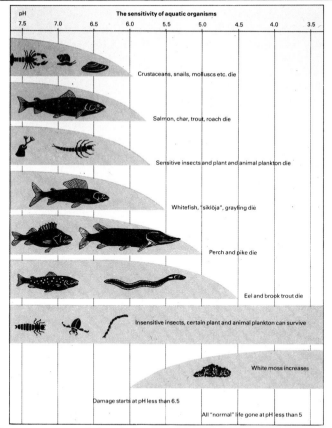

The sensitivity of aquatic organisms

pH								
7.5	7.0	6.5	6.0	5.5	5.0	4.5	4.0	3.5

Crustaceans, snails, molluscs etc. die

Salmon, char, trout, roach die

Sensitive insects and plant and animal plankton die

Whitefish, "siklöja", grayling die

Perch and pike die

Eel and brook trout die

Insensitive insects, certain plant and animal plankton can survive

White moss increases

Damage starts at pH less than 6.5

All "normal" life gone at pH less than 5

Tests have shown that some particles from high chimneys can travel as far as 2,000 kilometers (1,250 miles) in just three or four days. One huge, 400-meter chimney at a nickel and copper-smelting plant at Sudbury in Canada now emits as much sulphur dioxide every year – more than 500,000 tonnes – as all the natural sources in the world combined. (One natural source of sulphur dioxide, incidentally, is volcanoes but humans now out-generate all volcanoes by a factor of about 100.)

All rain is, in fact, slightly acid because it contains carbonic acid formed from the carbon dioxide in the atmosphere. (Those who study weathering effects on historic buildings are now wondering whether increased levels of carbon dioxide in the atmosphere – *see below* – might speed up the process.) Because of this, rain should theoretically have a pH value of 5.6, but in practice the

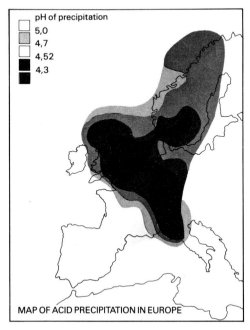

pH of precipitation
- 5,0
- 4,7
- 4,52
- 4,3

MAP OF ACID PRECIPITATION IN EUROPE

As acidity increases so different life forms are unable to exist. First the snails and molluscs go, then the fish begin to disappear – the eels last of all – leaving only a small number of insensitive insects and plants. Over 20,000 of Sweden's lakes have been acidified.

The map clearly shows the relationship between industry and acid rain. The long plume running SW– NE across Sweden illustrates the long-range transport of acid rain by the prevailing winds. The lower the pH value, the more acid the rain. That at the centre is nearly 10 times as acid as that round the fringes.

effect is largely neutralized by other substances in the atmosphere. (pH values are a measure of acidity in which a decrease of one, say from pH 6.0 to pH 5.0, denotes a ten-fold increase in acidity. A pH of 7.0 is neutral.) Since the industrial revolution, however, the acidity of rain has increased globally by between five and thirty times. In some places the rain is 1,000 times more acid than before. Technically, rain is called 'acid' if its pH value is actually less than 5.6. One of the most acid examples of rain ever recorded was at the start of a storm in Pitlochry in Scotland in 1974. The pH was 2.4 – which made the rain just as acid as vinegar.

In Sweden, where the effects of acid rain on lakes were first noticed, 20,000 lakes out of a total 90,000 have now been damaged by acid rain; 9,000 of them no longer contain any live fish at all. Norway is equally badly affected: 80 per cent of the lakes and streams in the southern half of the country are either 'dead' or 'critically ill'. In one part of southern Norway, there are no longer any live fish at all in lakes that cover a total area of 13,000 square kilometers (5,000 square miles). Both countries have kept careful records, and some lakes have shown a decline in pH values of as much as 1.8 units since the 1930s – which means they have become nearly a hundred times more acid in the past 50 years.

It was Sweden who first brought the problem to the attention of the international community – at the UN Conference on the Human Environment in Stockholm in 1972. Since then, acidification of lakes in Europe has been reported in Belgium, Denmark, Italy, the Netherlands and the United Kingdom.

In the eastern United States and Canada acid rain had been causing damage since the 1950s, but monitoring was not begun until the late 1970s. Now it is known that life has been damaged in thousands of Canadian lakes, and that fish have disappeared completely from more than 140 of them, mainly in Ontario. Scientists have predicted the imminent death of all fish in up to 48,000 lakes in Canada. When trout began to disappear from lakes in the Adirondack Mountains in New York State in the 1950s, it was believed that they were being eaten by perch – until the perch, in their turn, also began to die. Now the lakes there are so acidic that the area is known as a 'fish graveyard'. Nearly 90 per cent of the 217 Adirondack lakes that are above the 600-meter (2000-foot) contour have lost all their fish.

The ways in which acid rain affects aquatic life are not as obvious as might be thought. Acid rain mobilizes heavy metals such as cadmium and mercury in soils, rocks and sediments, which are then leached out by the rain and enter the surface waters. Acidified lakes have high levels of cadmium, lead, aluminium, manganese, zinc, copper and nickel. All these can kill living organisms if present in sufficient quantities, and it is thought that it is their presence, rather than the actual acidity of the water, that causes the death of fish and other life.

Once acidity in a lake begins to rise, reproduction of its inhabitants is impaired, calcium becomes depleted, aluminium leached from the surrounding soils builds up on fish gills and, at a pH of about 5.5, the smallest species disappear. If the pH then falls below 4.5, all fish die. As Erik Eckholm has put it: 'Lacking organic matter, the water of an acidified lake assumes the crystal clarity of a swimming pool – a deceptive beauty indeed.'

As well as its dramatic effects on aquatic life, acid rain acidifies soil and stunts the growth of plants. The local effects of sulphur concentrations are very evident in the United Kingdom, where they can be seen for hundreds of kilometers around the power stations that cause them. Lichens (mosses) are particularly sensitive to atmospheric sulphur dioxide and can be used as indicators of concentrations of the gas. According to Dr David Drew of Trinity College, Dublin, there is a 'lichen desert' around Newcastle-upon-Tyne which extends furthest on the town's downwind (south-eastern) side and covers 1,000 square kilometers (400 square miles).

Who Creates the Acid Rain?

The Tyneside area could well be one of the sources of the acid rain that falls on Scandinavia, much of which comes from the United Kingdom, East Germany and Poland. Canada suffers similarly at the hands of the United States: although the exchange of pollutants is a two-way process, Canada receives more than twice as much sulphur and eleven times more nitrous oxide than it sends south. These inequalities are an essential part of the problem, causing international bitterness whenever the issue is discussed. Because hardly any industrialized countries export the same amount of acid rain as they import, there is potential cause for dissension between many European countries (*see* table *below*).

The effects of acid rain on soils are greatest where the soils are thin and the buffering effects therefore minimal. In West Germany, scientists suspect that acid rain is the cause of the mysterious death of over 1,500 hectares (3,700 acres) of evergreen forest in Bavaria over the past five years. A further 530,000 hectares (1,300,000 acres) of forests (7.3 per cent of the total wooded area) are now also thought to be seriously affected. There are

Where the European sulphur ends up

Country	Emitted (tonnes of sulphur x 1000)	Deposited (tonnes of sulphur x 1000)
Albania	50	67
Austria	215	341
Belgium	404	161
Bulgaria	500	346
Czechoslovakia	1500	1301
Denmark	228	109
Finland	270	293
France	1800	1212
German Democratic Republic	2000	778
Germany, Federal Republic of	1815	1158
Greece	352	253
Hungary	750	467
Iceland	6	74
Ireland	87	65
Italy	2200	1132
Luxembourg	24	11
Netherlands	240	173
Norway	75	255
Poland	2150	1330
Portugal	84	73
Romania	100	797
Spain	1000	583
Sweden	275	472
Switzerland	58	141
United Kingdom	2560	847
Yugoslavia	1475	1093

All countries emit sulphur and have it deposited on them, but, as is often the case, some countries emit far more than they have deposited on them. This material ends up causing damage to other countries and raising difficult problems in international relations. Countries such as Belgium, East Germany and Britain are net exporters of this pollution; others, such as Romania, Sweden and Austria, are net importers.

signs that acid rain is also affecting the few forests that the United Kingdom has left. And scientists in North America are now beginning to wonder about their forests.

It has been estimated that the rate of forest growth declined by between two and seven per cent in southern Scandinavia and the north-eastern United States between 1950 and 1970 – but because terrestrial ecosystems are extremely complex, with so many natural changes going on at the same time, it is difficult to be certain that the decline is due to acid rain. Scientists expect the chemical balance of the soil to alter where nutrients are leached out by acid rain, and they are keeping a careful watch on the situation.

A recent study estimated that dealing with the minor effects of acid rain – repairing damage to bridges, fences and railings, for example – costs the world 1,450 million dollars a year. But this is only the tip of an iceberg of costs that have not yet really begun to be assessed. In West Germany acid rain is suspected of causing timber losses of 350 million dollars a year. The farmers there and in Scandinavia also blame acid rain for considerable agricultural damage: they must constantly lime their fields to counteract the increasing acidity, which tends to slow down the growth of all forms of crops, whether grass, cereals or trees. The Organisation for Economic Co-operation and Development has estimated that acid rain already costs its members 500 million dollars each year in lost crops.

In Poland, in the Katowice area, trains are limited to a speed of 40 kilometers per hour (25 miles per hour) because of corrosion of the rails by acid rain. Hundreds of historic buildings are crumbling away: the faces of stone statues and the gold roof of a sixteenth-century chapel in Cracow, Poland, have been severely damaged; Cologne Cathedral is dissolving; both the Parthenon and the Acropolis in Athens are being eaten away; and the facade of the Canadian Parliament building in Ottawa is turning black.

Drinking-water in parts of the Adirondacks in the United States contains unsafe levels of various heavy metals for which acid rain is believed to be responsible. There seems little doubt that ground water is being affected. Scientists from the Swedish Water and Air Pollution Research Institute, investigating complaints of corroded water pipes from householders in the province of Bohuslän on Sweden's west coast, discovered that the water had a pH of below 5.5 and contained high concentrations of copper and zinc, and some cadmium. The residents of the area were not surprised. They had already noticed the corrosion of their aluminium saucepans, the unpleasant taste of the water, diarrhoea in their young children, and even cases where blond hair turned temporarily green from being washed in water containing so much copper. At the same time, poultry farmers were complaining that their birds laid thin-shelled eggs after being exposed to acid ground water.

Cutting Down on Acid Rain
So what is to be done?

Attempts are now being made to neutralize acidified lakes by treating them with lime. In Sweden, the liming process was costing 4.5 million dollars a year at the end of the 1970s. By the summer of 1982 1,500 Swedish lakes had been limed, at a total cost of about 12.75 million dollars, and it was clear that the technique was at least partly effective in improving the state of acidified lakes. The Swedes reluctantly concluded that they would have to expand their liming campaign very considerably because 'it is going to be a considerable time before the emission of acidifying substances in Europe can be reduced to a level that will not entail large-scale acidification of lakes and watercourses'.

However, new air pollution laws and efforts to reduce urban air pollution have already reduced some sulphur emissions. The amount of sulphuric acid in European

Schloss Herten, the Ruhr, West Germany. The picture
on the left was taken in 1908, 206 years after the
statue was made. The picture on the right was taken
in 1968, clearly showing the devastating effect of air
pollution.

FLUIDISED BED COMBUSTION

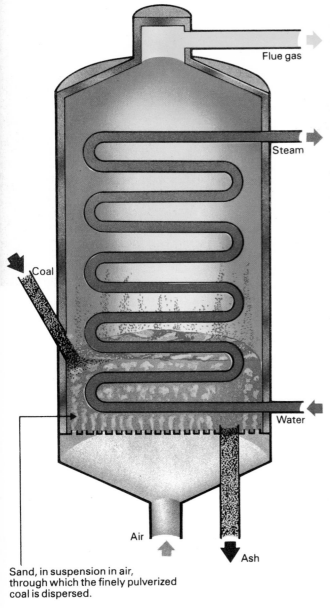

Flue gas

Steam

Coal

Water

Air

Ash

Sand, in suspension in air, through which the finely pulverized coal is dispersed.

rain levelled off slightly during the 1970s but nitric acid levels have continued to climb. In the United States, total sulphur emissions declined in the 1970s. However, in the face of energy policies that are becoming increasingly coal-based, maintaining current sulphur levels may be difficult, and reducing them even more difficult.

Despite its increasingly obvious effects, the issue of acid rain had not been publicly aired at the international level before it was brought up by Sweden at the United Nations Conference on the Human Environment. Between 1955 and 1965 Swedish and Norwegian weather stations had reported a rise in sulphate and nitrate concentration and in acidity in rainwater. A Swedish case study presented to the conference suggested that transnational action was urgently needed.

In 1979, the United Nations Economic Commission for Europe drew up a Convention on Long Range Transboundary Air Pollution, obliging parties to it to make every effort to reduce air pollution, especially the discharge of pollutants carried by winds across frontiers. A further stipulation was that if any nation planned activities likely to increase air pollution, it should be obliged to notify and consult countries downwind of it. Of the Commission's 34 members, 31 signed the Convention, which eventually came into force in March 1983. There were great hopes that it would be instrumental in reducing sulphur dioxide emissions.

Certainly West Germany was keen to play its part. After German scientists laid the blame on acid rain for the state of their spruce and fir forests, the government pledged itself in 1983 to achieve a 50 per cent reduction in its emissions of sulphur and nitrogen oxides over the following ten years. It intended to achieve this by using flue gas desulphurization systems in both old and new power stations, and by regulations to reduce motor vehicle exhaust gases.

Measures like these will help but

Fluidised-bed combustion, an advanced technology for burning coal, is one of a number of technologies available for reducing air pollution. It can eliminate most of the sulphur and nitrogen oxide emission from coal use. Several fluidised-bed combustors are already in use around the world.

they are not cheap. Countries such as the UK and the United States, which are among the major exporters of acid rain pollution, still maintain that more research is needed into exactly how acid rain is caused, and what the best solutions are. Whether the newly ratified Convention on Long Range Transboundary Air Pollution will speed things up remains to be seen.

Since 1980, the United States and Canada have been trying to agree on how to combat transboundary air pollution. By 1982 a draft agreement was ready. But there the talks stuck. In 1983 the Canadians maintained that the evidence proved a need to fix an immediate limit on sulphur dioxide emissions (they suggested 20 kilograms per hectare [18 pounds per acre] per year), but the United States Government did not agree that the information then available was enough to justify such a step. Wrangles like these will probably become more and more common as nations struggle to implement international legislation against acid rain.

But the issue is no longer the preserve only of governments, experts and scientists. In the summer of 1982, a group of young Americans and Canadians travelled through Ontario and the north-eastern United States in a converted white bus called 'The Acid Rain Caravan', publicizing the issue with the help of literature, slide shows, T-shirts and badges. And the International Youth Federation of Environmental Studies and Conservation proclaimed an International Acid Rain Week in April 1983, with the intention of raising public awareness. Acid rain may be turning into another environmental issue on which people are reluctant to let their governments set the pace.

The 'Hair-Spray' Threat
The Earth is protected from the Sun's ultraviolet radiation by a layer of ozone high up in the stratosphere. Without it, human life could not exist on Earth. And if,

The humble hair-spray may pose a serious threat to the ozone layer which protects us from the harmful effects of the sun's ultra-violet radiation.

as now seems possible, human action on the Earth below were to damage this ozone layer, the increased levels of ultraviolet radiation that would be received on the planet's surface could do a great deal of damage; of the known effects, there would be an increase in certain types of skin cancer, in certain types of eye disease, and changes in the rates of plant growth. There might well be other effects of which we are not yet aware.

The ozone (O_3) – three oxygen atoms joined together in each molecule – is produced from ordinary oxygen (O_2) – two oxygen atoms joined together in each molecule – and decomposed when ultraviolet radiation is absorbed, or by chemical reactions with various atmospheric trace gases. What actually happens is that ozone is continuously produced and continuously decomposed.

The quantity in the atmosphere at any one time is comparatively tiny: less than 3,300 million tonnes (compared with, for example, 3,865 million million tonnes of nitrogen). However, this sparseness means that it is relatively vulnerable to outside disturbance.

Several human activities could damage the ozone layer – and some may already have done so. The increased use of nitrogen fertilizers could lead to ozone depletion. So could increasing the numbers of supersonic aircraft (SSTs) operating in the stratosphere – a prospect which no longer seems very likely, but which caused considerable worry when Concorde was being developed. The third threat to the ozone layer, and the one that is now taken most seriously, comes from substances called chlorofluorocarbons (CFCs), which are used as refrigerants, fire extinguishers and as the propellants in

Concorde: economics, rather than environmental opposition, have reduced the potential threat to the environment.

aerosol spray cans.

One of the things that contribute to the natural destruction of ozone in the stratosphere is nitrous oxide (laughing gas, N_2O). If the use of nitrogen fertilizers continues to increase, more nitrous oxide will be released into the atmosphere, and more will get into the stratosphere. Scientists suspect that this will cause ozone depletion; but the effects are complicated, and recent estimates of how much depletion would result, for instance, from doubling the concentration of nitrous oxide in the atmosphere have varied from as much as 20 per cent to the current estimate of somewhere between two and six per cent.

The second source of worry about the ozone layer came in the late 1960s, when it seemed that SSTs would revolutionize the transport industry. Because SSTs fly at stratospheric levels, they dump oxides of nitrogen and water vapour directly into the stratosphere's chemical reaction chains – with just what results, nobody is quite sure. In the late 1960s, it was predicted that there would be more than 500 civil SSTs operating daily in the stratosphere by the early 1980s. In fact, there are now only about a dozen Concordes in service and a few Soviet SSTs confined to internal flights. Their effect is not likely to be important. It could become so, however, if supersonic flight were ever to become more common.

The chemicals known as CFCs are, for the moment, by far the most serious threat. The innocuous-looking aerosol spray, if used in sufficient quantity over the next few decades, could almost certainly lead to widespread changes in the state of our planet.

In the 1930s, researchers were looking for refrigerants which boiled between 0 and −40 degrees Centigrade (32 and −40 degrees Fahrenheit), which were chemically stable, non-poisonous, cheap, easy to produce and easy to store. They found a group of chlorofluorocarbons with the right properties, and two of them were marketed under the names Freon-11 and Freon-12. They were used in refrigerators and for fire extinguishers, but their big moment came in 1950 when the first aerosol can of Freon-propelled hair spray was put on the market. Today, aerosol sprays consume about one million tonnes of CFCs a year.

The chemical stability of these CFCs means that they are not decomposed by chemical or biological processes and they gradually accumulate in the atmosphere. They may contribute to the warming of the atmosphere caused by the increase in carbon dioxide (*see below*). In the stratosphere, however, they are decomposed by ultraviolet radiation, creating chlorine atoms in the process. The chlorine thus formed destroys ozone, it is suspected, in a way which could seriously interfere with the balance that is normally maintained between the rates of production and decomposition of ozone in the atmosphere.

In 1981 the United Nations Environment Programme (UNEP) estimated that if releases of Freon-11 and Freon-12 continued at the 1977 rate, the stratospheric ozone layer would eventually be depleted by about 10 per cent. In 1979, the US National Academy of Sciences put the figure at 16.5 per cent. These figures were, however, only estimates and did not, in any case, take account of additional effects which might be contributed by emissions of other CFCs or other similar chemicals such as methylochloroform and carbon tetrachloride.

The Chemical Manufacturers Association reported in 1980 that world production of Freon-11 and Freon-12 had fallen, between 1974 and 1979, by 17 per cent. UNEP asked nations to agree to regulate production and use of CFCs, and the United States and Sweden had begun to do so by 1982. Preliminary discussions organized by UNEP on an international convention to reduce emissions of CFCs began in 1981.

If the ozone layer is being depleted, a one per cent reduction could cause anything up to a 3 per cent increase in the amount of ultraviolet radiation reaching the surface of the Earth. The US National Academy of Sciences has also calculated that each one per cent of ozone depletion would cause an increase in the incidence of skin cancer of about 2 per cent. Thus, if the ozone layer were to be reduced by 5 per cent, there would be 40,000 more cases of skin cancer each year in the United States alone.

Figures such as these are estimates based on the best current scientific understanding of what is going on in the atmosphere. However, they are only estimates, and not all authorities agree on how best to 'model' the chemical reactions taking place in the ozone layer. Furthermore, it is difficult to be sure whether any damage has yet been done,

mainly because records have not been kept for very long and current technology is not capable of detecting the tiny changes which may be occurring on a day-to-day or month-to-month basis. If the model used by the UNEP committee is correct, a total ozone depletion of about one per cent should already have occurred. But it has not proved possible to measure this, nor do levels of ultraviolet radiation reaching the Earth appear to have altered – but monitoring began only recently.

The problem of ozone depletion is likely to be typical of more and more environmental questions facing mankind. Because the effects of depletion are potentially so serious, it may be necessary to take action to avoid depleting the ozone layer even before it can be determined whether or not it is actually being depleted. Although the effects on humans are serious enough, a more major, longterm worry is what increased ultraviolet radiation might do to the general ecology of the planet.

Ultraviolet radiation has its most virulent effect on small, single-cell organisms, on insects, on plankton and on microbes. It certainly affects the speed at which most plants grow. It is therefore quite possible that changes in the ultraviolet radiation at the Earth's surface would have a profound and essentially unpredictable effect on agricultural productivity and on ecology in general. Anything which alters the balance between species on the Earth is to be treated with caution. Under the circumstances, it seems only sensible to take every precaution possible. Unfortunately, while hair sprays may be relatively easy to ban, there are other ways in which human activity is having large-scale effects on the planet's equilibrium which may be far more difficult to correct.

By far the most important of these could prove to be the effect of this century's much increased amount of carbon dioxide in the atmosphere. The possible climatic results of the consequent increase in the so-called 'greenhouse effect' are so far-reaching that this topic is discussed in Chapter Ten: Will the Climate Change?

Implications for the Future of Our Air

The atmosphere, although it affects everyone, belongs to nobody. No one has sole responsibility for it. This makes it more difficult to protect and easier to pollute. It also means that even those who do not pollute it suffer from the pollution of others. The problem, augmented by human ignorance, of how best to treat the atmosphere is compounded by the fact that it is an international resource, and to achieve anything nations must first agree on a course of action.

Yet the mechanisms that exist to reach such agreements are still woefully inadequate. It is true that the onset of environmental consciousness on the planet has helped improve them somewhat. The United Nations Environment Programme is certainly the only international body that has yet managed to persuade all the Mediterranean nations, many of whom are now mutually hostile, to sit down at the same table and prepare a plan to clean up their common sea. But in other areas – and notably in the field of acid rain – international accord is not so easy. Nor is there any magic solution. In the end, technology can provide no alternative to hard political debate.

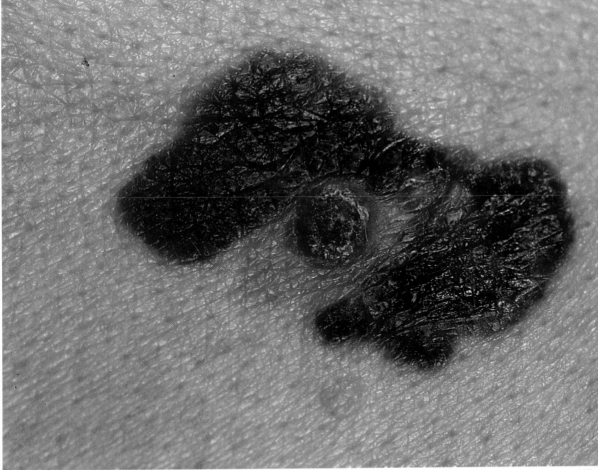

Skin cancers like these will increase if there is a significant decrease in the ozone layer.

But this is not the only problem. Even if mechanisms of reaching international agreement more easily were at hand, the knowledge on which to base such agreements is very often missing. Our ability to befoul our planet, on a global scale, has surpassed our ability to understand the details of what we are actually doing. It is, as more than one commentator has put it, as though we were conducting a giant geophysical experiment in which the Earth was the subject and its inhabitants the victims.

In the face of all these threats, some implications are obvious. One is that we should keep open as many doors to the future as possible. Our knowledge of the mechanics of carbon dioxide absorption into the great chemical cycles of the Earth (see Chapter Ten) is still too slim for us to be able to make accurate predictions. It is therefore prudent not to make decisions that commit the human race to only one course of action. For example, because of the dangers of nuclear power, a decision might be made to abandon nuclear reactors and to develop coal-based reactors to last for, say, the next 200 years by which time nuclear fusion should be available.

Such a decision, however, could have disastrous consequences. And, with nuclear power abandoned, there would be no easy way to turn back. If, as now seems highly likely, mankind has reached a critical stage in its evolution on the planet, we must begin to hedge our bets and keep all our options open.

One reason for this is that the time-scales involved have become extremely awkward. A point has been reached at which urgent action is needed to prevent further deterioration. Yet the knowledge needed to take that action is not necessarily yet available. We need more time – but we may not have it.

Robin Clarke

Tokyo policeman wearing smog mask — will our children have to learn to live with pollution?

Chapter 8

The City Environment

As cities grow in size and population, and as an ever higher proportion of the world's people come to live in urban areas, two major issues link environment and cities. The first is what one can term cities' external impact: the extent to which they draw on resources such as land, fossil fuels and fresh water and create environmental problems for the region, nation and ultimately the planet. The second is the environment that cities provide for their inhabitants.

In this chapter, 'cities' are taken to mean urban settlements that play an important role in their nation's economy. No range of population size can be specified since an important city in a poor, small African nation would have the same population as a relatively small city in the West or in, say, Brazil. There are no internationally agreed criteria for distinguishing between 'urban' settlements and 'rural' settlements. Every nation has its own criteria and in some

Apartment block in Tokyo — the second largest urban area in the world. Currently, 20 per cent of the global population live in cities of over 100,000. By the year 2000 this will have risen to 34 per cent.

149

nations these can mean that settlements with less than 1,000 inhabitants are 'urban' whereas in others, settlements of 5,000 or even 10,000 inhabitants are 'rural'. This greatly limits the accuracy of international comparisons of 'urbanization levels'.

Urban Growth

During the last 100 years, the world has experienced a growth in urban population and a concentration of population in large cities which, as a general trend, is unprecedented in human history. Behind this trend has been the growth and development of an ever more integrated and urban based world economy. As national economies and people's livelihoods become less and less based on agriculture, so the city and its wider metropolitan area replace the village, the farm or the seasonal wanderings of the pastoralist as the main location for human activities. It is within an increasingly interlinked network of cities that world production and trade takes place. The workforce has to live close by their work, so residential areas are close to production centres. Only a tiny, rich elite can work in such centres and commute from houses set in rural peace.

There has also been a parallel process of settlement in large areas which only 40 or 50 years ago were either uninhabited or only sparsely populated. The search for new farmland and pasture, for timber and non-renewable mineral resources has underpinned the rapid colonization of new land, both through government programmes and through spontaneous, uncontrolled population movements. The eastern regions of Peru, Bolivia and Ecuador, Amazonia, southern Nepal, and large areas in Indonesia provide examples of this. Some of the more rapidly growing cities are those which grew up as service and processing centres for these newly settled areas.

In 1920, the world's entire urban population was only some 360 million. By 1950, it had essentially

doubled. It had doubled again by the early seventies, and is likely to double again by the mid-1990s. By the end of the century, slightly more than half the planet's population is likely to live in urban areas. Around two thirds are likely to be in cities of 100,000 or more inhabitants, while more than two fifths will probably be in cities of a million or more inhabitants.

There are enormous national and regional differences both in the proportion of people living in urban areas (the so-called urbanization level) and urban growth trends. Northern and western Europe, Japan, temperate South America, North America, Australia and New Zealand all have between 70 and 90 per cent of their people in urban areas. Many of their larger cities are no longer growing

Guayaquil, Ecuador. Most of the anticipated urban growth between now and the end of the century will take place in existing cities in the developing countries. This will often entail expansion on to unsuitable land, such as the flood plain shown here.

are major regional differences within nations both in levels of urbanization and in rates of urban growth. For example, in Mexico some of the more prosperous and developed states have three quarters or more of their population in urban areas while the poorer states on the south-west coast have predominantly rural populations. While some states in India have more than 30 per cent of their inhabitants in urban centres, others have less than 10 per cent. Again it is generally those areas within nations with larger and more diversified roles within the world economy which have urbanized most rapidly over the last thirty years. It is generally the poorest regions that remain the least urbanized.

While rapid urbanization first became a general trend in Europe, North America and Australasia, it is the three Third World regions – Africa, Asia and Latin America – which are now spectacularly increasing their share of the world's urban inhabitants. China has the world's largest national urban population, even though most of its people are still in rural areas. India's population is mostly in villages but its urban population is now larger than that of the United States. Brazil has close to 100 million urban residents. The Third World's urban population is likely to exceed 2,000 million by the year 2000; it will account for more than two thirds of the entire world's urban population. As recently as 1950, Cairo was Africa's only city with more than one million inhabitants; by 1980 there were nineteen African 'million cities', and by the year 2000, there will probably be more than sixty.

The External Impact of Cities
Cities and their wider metropolitan areas are the great centres of production and transportation because the production of goods, services and information is concentrated within them or nearby. They are the centres of political and economic power. The major decisions regarding the present and future worldwide distribution

rapidly; indeed, some are declining in population. At the other extreme, East Africa's urbanization level is only around 20 per cent while that of South Asia and China is around 25 per cent. Generalizations are difficult, because between nations there are major differences between urbanization levels and urban trends. But as a rough rule of thumb, the higher a nation's per capita gross national product (GNP) and the larger its role within the world economy, the higher the percentage of its population living in urban areas. The poorest nations, whose economies play little role in the world economy, are generally those with the lowest levels of urbanization.

But looking at urban trends at the national level obscures the fact that there

Growth in urban population
1950-75

- over 400%
- 301 — 400%
- 201 — 300%
- 101 — 200%
- 51 — 100%
- 50% and below
- urban decline
- data not available

As population and economic activity have increased,
so more of us have become city dwellers. In the
industrialized countries this trend is slowing down
and may even be going into reverse in some cases. In
the developing countries, however, urbanization is
proceeding faster than governments can handle.

of productive investment are made in relatively few cities. And since urban growth trends tend to reflect economic growth trends, many nations' or regions' economic and thus urban growth prospects are much influenced by decisions taken outside their boundaries and outside their control.

Cities and metropolitan areas are also the major consumption centres. They demand a continual large input of resources to meet production and consumption needs – water, energy, and all the goods and materials that their populations and their enterprises require. They are also great centres for resource degradation. Vast volumes of water are needed, so cities draw on fresh water resources, often from huge areas around them, and usually return waste water to rivers, lakes or the sea at a far lower quality than that originally supplied (*see* Chapter Six). Cities pollute the air with, among other things, emissions from industry and internal combustion engines (*see* Chapter Seven). Their physical growth often takes up high quality agricultural land. Their functioning demands large inputs of fossil fuels or other energy sources for transport, for power, for machines and for heating or cooling buildings. Just one highrise building, whether in New York, Sao Paulo, Delhi or Lagos, often demands more electricity to keep it functioning than that needed by many villages or even a small city.

There is growing concern that large cities, especially the more rapidly growing ones (most of which are in the Third World) will reach a size where the beneficence of nature, in terms of providing and then purifying air and water and yielding cheap and easily exploitable energy sources, will be insufficient. This concern is heightened by looking at projections up to the year 2000 and beyond (*see* page 155). Today, there are numerous examples of environmental stress that seem likely to multiply as cities grow ever larger.

Providing metropolitan areas like Mexico City, Sao Paulo or Los Angeles with adequate water supplies already presents huge problems. Cairo, Lima and Bogota, in common with so many other cities, continue to spread across some of their nation's most fertile agricultural land.

Then, in many Third World countries, there is the problem of increasing deforestation in swathes around cities. As widely used oil-based fuels like kerosene (paraffin) have risen in price and as more people find themselves unable to afford them (or poor distribution systems make their availability erratic), the demand for fuelwood and charcoal has also risen (see Chapter Four). Many cities are major centres for the consumption of firewood; the poorer the city's inhabitants, the more likely they are to have to rely on firewood for fuel. Rising fuel prices make the sale of firewood commercially attractive, and rapid deforestation around population centres often goes hand in hand with high profits for those engaged in the firewood

trade.

Finally, many large cities are outgrowing their original natural site. For instance, when Rio de Janeiro was founded in what is one of the world's most beautiful natural sites, no one could foresee that it would grow to be a multi-million inhabitant metropolitan area. Population growth, combined with uncontrolled low-density sprawl on cities' peripheries, causes huge increases in the costs of providing newly developed areas with roads, water and sewerage. The haphazard growth of cities often destroys areas of great natural beauty. While this is more evident in cities which grew up in particularly magnificent sites, it is happening in virtually all growing cities with loss of much-needed open space and recreational areas.

These external pressures are enormous and usually dominate the debate about city environments. They are so obviously in evidence, especially in the major centres of heavy industry. For

154

Sao Paulo, Brazil. Air pollution, inadequate water and sewage services, poor housing, expensive transport and energy are just some of the problems facing the rapidly growing cities of the South.

Figure 1. The world's ten largest urban areas (population in millions)

1950		
1.	New York — NE New Jersey	12.3
2.	London	10.4
3.	Rhine—Ruhr	6.9
4.	Tokyo—Yokohama	6.7
5.	Shanghai	5.8
6.	Paris	5.5
7.	Gran Buenos Aires	5.3
8.	Chicago — NW Indiana	4.9
9.	Moscow	4.8
10.	Calcutta	4.6

1980		
1.	New York — NE New Jersey	20.2
2.	Tokyo—Yokohama	20.0
3.	Mexico City	15.0
4.	Shanghai	14.3
5.	Sao Paulo	13.5
6.	Los Angeles—Long Beach	11.6
7.	Peking	11.4
8.	Rio de Janeiro	10.7
9.	Gran Buenos Aires	10.1
10.	London	10.0

2000		
1.	Mexico City	31.0
2.	Sao Paulo	25.8
3.	Shanghai	23.7
4.	Tokyo—Yokohama	23.7
5.	New York — NE New Jersey	22.4
6.	Peking	20.9
7.	Rio de Janeiro	19.0
8.	Greater Bombay	16.8
9.	Calcutta	16.4
10.	Jakarta	15.7

example, Mexico City's air pollution is visible to all visitors. But the last fifteen years have shown that polluting emissions *can* be controlled, given the political will of leaders, and the appropriate legislation and its enforcement. New industrial plant designs can recover and re-use what was previously dumped in rivers or into the air. The efficiency with which fossil fuels are used can be greatly improved. Furthermore, many of the more modern and rapidly expanding industries (such as electronics) do not demand the massive consumption of natural resources of, say, the steel industry. In the West, the employment base of cities is increasingly in service industries rather than heavy industry. While worrying setbacks to improving environmental quality can be seen, the technical and financial ability to tackle the major environmental problems is ultimately available.

The Internal Impact of Cities

But these external pressures are minor when compared to the problems of the internal environment, that is, the environment cities provide for their inhabitants. And it is in Third World cities that this problem is concentrated. This environment is not so much despoiled by noise, industrial pollution or automobile exhausts, although these play their part. It is primarily despoiled by poverty and by the refusal (or inability) of city and national governments to tackle its underlying causes.

In terms of environment, most major Third World cities are really *two* cities: the city of the elite where Western standards are evident, and the largely self-built city of the poor. The very visible difference between those living in low-density, well-serviced residential areas and those living in illegal shacks built of scrap materials, simply reflects the more fundamental difference between them in income and (thus) in political influence. A third or more of the inhabitants of most Third World cities live among the world's most

155

In 1950 seven out of the top ten cities were in the industrialized world; now only four of them are; by the year 2000, only two will remain.

degraded environments. In terms of services, a lack of readily available drinking water, sewerage connections (or other systems to dispose of human wastes), garbage collection and health services ensures that many diseases are endemic – diarrhoea, dysenteries, typhoid and food-poisoning among them.

In terms of houses, crowded, cramped conditions mean that communicable diseases like tuberculosis flourish – usually aided by the low resistance among the inhabitants due to malnutrition. So too do household accidents, perhaps not surprisingly when four, five or more persons live in one room and there is little chance of giving the occupants (especially children) protection from fires, stoves and kerosene

heaters. Inadequate incomes combine with inadequate houses and services to provide totally degraded housing environments.

In terms of site, people are forced to live on land unfit for human habitation. The fact that no conventional dwellings exist at an affordable price (either to buy or to rent) and that no serviced house plots on which to build are available, means that people are forced to live on illegally-occupied or subdivided land. Often, the only sites on which the poor stand some chance of not being forcefully ejected are on land ill suited to commercial development, such as ravines or slopes subject to landslides (as in Rio de Janeiro and Caracas) or floodplains or tidal basins (as in Guayaquil, Lagos and

Rio de Janeiro, Brazil. The stark contrast between the affluent centres and poverty-stricken outskirts is typical of many Third-World cities.

Shanty town, Lagos, Nigeria.

countless other cities). Since these settlements are 'illegal', very rarely do public authorities provide them with water, sanitation, drainage, public transport and basic social services.

Shanty Towns

These settlements of the poor are not isolated phenomena. They have become the norm. They exist and house tens or even hundreds of thousands of households in virtually every major Third World city. These illegal, largely self-built settlements remain the major location of new housing construction in most Third World cities. Of course, in many districts in Western cities, the inhabitants also suffer from a very poor living environment. But at least the population has piped

water and sanitation, some access to social services and usually some economic support. One child in three does not die before the age of five as happens in most Third World slums and shanty towns.

No easy solution exists. Slowing migration to the cities would make the problem more manageable. But this would require improved economic opportunities in other locations. The concentration of urban population in a few (or indeed one or two) large cities within nations is a response to the concentration of private and public investment there over the last 30 years, and also to the forces in rural areas which dispossess peasants and landless workers of their livelihood as described in several earlier chapters.

In the immediate future, it is unlikely that these trends will change. Financial returns on private capital remain higher and more certain in major cities. Only rarely do firms located in cities pay the costs their operations generate for the community in terms of increased migration and increased demand for land and water, services and infrastructure. Even if the flow of people could be diverted to small cities or encouraged to stay in agriculture because of better economic opportunities there, most Third World city populations would still grow at 2.5 per cent or more a year from natural increase. Indeed, in many cities, natural increase already plays a more significant role in population growth than migration.

The actions needed to tackle these 'internal' environmental problems are well known, although little implemented. The first factor is obvious – that unless the poor secure an adequate, stable economic base, no strategy to improve their environment will have great impact. Without this, and without more healthy living environments, birth rates are also unlikely to fall substantially. The first action – easier said than done – then must be to create a more stable economic base. The second action must be a national programme that underpins and supports poor communities' efforts to rebuild and improve their own habitat. This is often called the 'self-help' approach, although it still implies a major government programme for the provision of water, sanitation, drainage, garbage removal, primary health care, education (including campaigns to improve knowledge of personal hygiene and health care), public transport, cheap loans, legalization of tenure for illegal settlements' inhabitants and provision of cheap, serviced housing plots in growing cities.

Some of the more successful projects of the last ten to twenty years have been based on such an approach – working with slum or squatter communities in determining their needs, improving services and providing cheap serviced plots for new house construction. Projects where the people themselves played a major role in designing and implementing the projects, and in managing house construction or house upgrading, have usually worked far better than government projects to build public housing 'for the poor'. Public housing usually turns out to be too expensive for the poor to afford (even when subsidized), poorly located (which means extra costs in terms of time and money spent travelling to work) and poorly designed in terms of households' needs for space and the ability to improve and extend their own dwelling. The growing internationalization of building design and technology can be seen in the drab uniformity of conventional housing projects, supposedly for the poor, built all over the world. While such housing has rarely meshed with the poor's needs in the West, it has proved to be even less appropriate to the poor's needs in the Third World.

The trouble is that the 'self-help' rather than the 'public housing' approach can be used as an excuse by governments to do nothing. Many poor households have very little time available in which to build or improve their own houses. Their survival often depends on more than one of their members working extremely long hours. So government initiatives must

come to understand the range of housing needs among the poor. But what is needed most is government commitment, and the institutional structure to improve provision for these diverse housing needs, on a scale sufficient to have an impact. For national government, this means more support for providing basic social and physical services to the entire population. It implies better co-ordination between health, education, planning, housing and public works ministries or agencies operating at different administrative levels for concerted attacks on the environment of poverty. It implies careful consideration of the needs of all groups in society, including the special needs of children and youth. It implies more national support for city governments, since these mostly lack the funds, power, and skilled manpower to mount such a campaign.

Urban land markets provide an example of needed action at the level of cities and even nations. At present, the largely uncontrolled urban land markets price a third or more of most major cities' population out of legal housing. Only by public action to ensure a supply of legal, serviced plots which match the purses and priorities of lower-income groups can there be any slowing of environmental deterioration and any control of a city's

Top: self-help schemes, such as this one in Nicaragua, often offer the most hope of housing the poor and dispossessed.

physical growth. This demands action at city level, backed by national legislation, and financial and technical support. Uncontrolled market forces have *never* provided an adequate city environment for the poor.

Do not underestimate the difficulties. Land speculators and associated financial interests will oppose many of these changes, sometimes aggressively. But with or without public support, the so-called 'informal sector' will dominate housing construction in most major Third World cities. The effectiveness of 'self-help' can be measured by the rate at which the environment of poor households and communities improves. At present, many

Massive government expenditure will be necessary to provide basic services such as this water pipe in Delhi. Sanitation, garbage collection, public transport, health and education facilities are the minimum criteria of civilized urban life.

The appalling environment of poverty is at present all too obvious. It is most visible in Third World cities, although it is just as prevalent in rural areas. Extrapolating present trends suggests that what is already the world's greatest environmental catastrophe in terms of its impact on human lives can only get worse. But trend is not destiny. Few regions in the world lack the resources needed for a better environment to be built for the lower income groups. Given political will within governments, given support from the international community, and given policies that truly get to the root of poverty, this environmental catastrophe can still be tackled. It would be a notable triumph for 'environmentalists' and yet another example of the unity between sound environmental practice and sound development if the condition the poor live in – whether in large cities, towns or rural areas – were recognized as the environmental priority of the 1980s and 1990s, and we as an international community acted accordingly.

(This chapter is an expanded version of an article that first appeared in *People* magazine, and in *IUCN Bulletin*.)

**Jorge E. Hardoy and
David Satterthwaite**

governments' claim to increasingly support 'self-help' is no more than an attempt to legitimize their lack of action, and the shanty towns remain in a shocking state.

The Future
To predict what the world's cities will look like in the year 2000 is a dangerous task.

New office block in Bombay, India. The uncontrolled operation of urban land markets often means the domination of compassion by commerce. As the commercial activities of the city expand, so the poor are pushed further and further away onto more and more marginal land.

A concrete sewer pipe in Lagos houses the homeless
poor. Such will be home for tens of millions of people,
unless governments respond urgently to the growing
pressures of urbanization.

THE TRIBAL PEOPLES

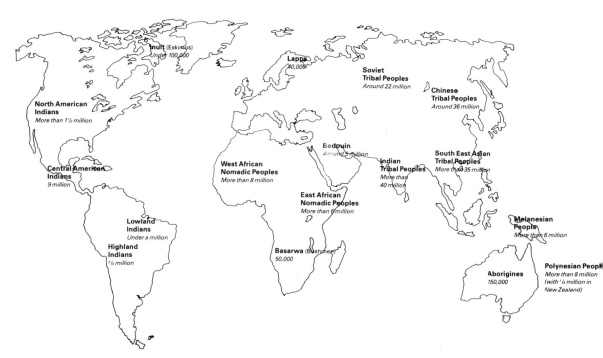

Inuit (Eskimos)
Under 100,000

Lapps
40,000

Soviet
Tribal Peoples
Around 22 million

Chinese
Tribal Peoples
Around 36 million

North American
Indians
More than 1½ million

Bedouin
Around 5 million

West African
Nomadic Peoples
More than 8 million

Indian
Tribal Peoples
*More than
40 million*

South East Asian
Tribal Peoples
More than 35 million

Central American
Indians
9 million

East African
Nomadic Peoples
More than 6 million

Melanesian
People
More than 6 million

Lowland
Indians
Under a million

Highland
Indians
¾ million

Basarwa (Bushmen)
50,000

Aborigines
150,000

Polynesian People
*More than 8 million
(with ⅓ million in
New Zealand)*

groups amounting to somewhere between two and six million people only five centuries ago, and perhaps one million people as recently as the year 1900, there are now only half as many such groups, with a total below 50,000 persons.

Their Loss Will Be Our Loss

Will it in fact matter if these curiosities become civilized out of existence? In the view of some observers, it will matter a great deal. The elimination of their relict cultures will mean the irreversible loss of 'knowledge banks' whose value can hardly be estimated.

In Northern Thailand, the Lua tribe, for instance, grow as many as 75 different food crops, 23 medicinal crops, twenty plants for ceremonial and decorative purposes, and seven plants for weaving and dyes. Other tribes in Indo–Burma utilize well over 100 plants for drugs, and almost 300 for foods. Through this extraordinary range of plant products,

these forest-dwelling tribes achieve a partial mimicry of the ecological diversity of tropical rain forests, with all the environmental stability that diversity fosters – by contrast with monoculture farms favoured by cultivators of the modern type. Among these dozens of obscure crops are new sources of food that could serve to improve nutritional levels in many parts of the world. For example, it is the forest peoples of Papua New Guinea who have provided us with knowledge about the winged bean, a protein-rich plant that they have cultivated in their 'garden farms' for centuries, without the plant becoming known to the outside world. Following its 'discovery' by the US National Academy of Sciences, this protein-rich plant is now being cultivated in more than 50 countries of the developing tropics.

In similar style, the Tsembaga people of New Guinea raise 264 crop varieties, frequently growing as many as 50 different

There are about 170 million tribal peoples in the world whose way of life, culture and often very existence is threatened by mainstream development.

kinds in a single garden patch. In the Philippines, the Hanunoo people are acquainted with 430 crops, commonly growing 40 of them in a two-acre plot. Who knows what new foods are available among their centuries-old farming customs, offering the potential to improve the daily diets of millions of people throughout the developing tropics?

Moreover, certain of these tribal peoples have learned how to sustain themselves with 'wild' foods from their environs. In the Tabasco State of south-eastern Mexico, the Lacandones gather at least 100 species of fruits and other wild foods from the surrounding forests, plus protein in the form of twenty kinds of fish, six of turtle, three of frog, and two each of crab, crayfish, crocodile and snail. This sustainable use of the forest's bounty is paralleled in south-east Asia by the Sav aran Kenyah tribe, who make up almost two thirds of their diets by gathering plants and hunting animals and

fish – and thereby offering us insights into entirely new sources of food, especially of animal protein. In similar style again, the Trans-Gogol people of New Guinea utilize 42 bird species, sixteen mammals, eight reptiles and two frogs. These same people also draw on forest wildlife for materials to use as garments and decorations, weapons and tools, food containers, cordage, house-building materials, musical instruments, medicines, narcotics, stimulants and intoxicants.

The self-reliant tribals turn hardly at all to the outside world to keep themselves going. For their agriculture, they supply their own variations of fertilizers and pesticides: what a boon it would be for the world's farmers if we could devise cheaper ways of supplying nutrients to our crops, and keeping insect pests away from them! Tribal people produce fertilizer through organic mulches, green manure, crop residues, nitrogen-fixing legumes, livestock wastes,

Witch doctor in Dahomey. As scientists begin to investigate folklore systematically, they discover that it contains a wealth of practical information on useful drugs. Much of this knowledge could be lost if tribal cultures are allowed to disappear.

The Tuareg of the Sahara have developed patterns of social organization and political accommodation that have much to offer the developed world's social scientists.

farm debris, household refuse and general garbage. They also reduce their needs for fertilizer by utilizing combinations of crops that help maintain soil fertility. Were they to farm after the manner of the outside world, through planting mainly annual crops plus only a few semi-perennial crops, they would quickly exhaust the soil of its fertility. But when they integrate short-term crops with at least as many perennial crops, they reduce or even prevent the impoverishment of the soil.

In addition, some 'garden farmers' exploit innovative sources of fertilizer. An obvious example is bird guano from the forest. Still more exotic, while equally effective, is the fertilizer utilized by farmers in the Oaxaca Valley in Mexico; this is organic debris ejected from the fungus gardens of atta ants, which resembles fine sawdust and is extremely rich in nutrients.

As for pesticides, the farmer of the tropics lacks that great pesticide of temperate zones known as winter. But he does not need to buy expensive pesticides from giant chemical corporations; he employs his own biological controls. In multi-crop fields, the variety of plants serves to hold down pest populations by limiting their food and living space, as well as by fostering natural enemies such as insect predators and parasites. The inventive farmer can also call upon the natural pesticides contained in the tissues of certain plants. If, for example, he plants maize and peanuts in alternate rows, he exploits the pest-inhibiting biocompounds of peanuts in order to reduce the depredations of burrowing insects that often infect his corn crop.

All in all, then, traditional farmers of 'primitive' tribal communities demonstrate a sound way for subsistence peasants to make out with meagre resources. If their agricultural strategies could be expanded and made available for many more people, they could constitute precisely the type of agro-ecosystems that we need to meet the challenge of sustainable farming in tropical lands. In fact, we could even anticipate an improvement on the Green Revolution; it might be known perhaps, by virtue of the myriad plant and animal species involved, as the Gene Revolution.

'Primitive' Medicine
So much for the potential contributions of tribals to modern agriculture. As for medicine, the Amerindians of Amazonia are acquainted with 750 plant species

Pygmies of the Ituri forest – a people that are threatened by the destruction of the rain forest in which they live.

Eskimo. The development of Arctic oil resources has brought great wealth to some of the Eskimo peoples, but at the price of loss of their traditional way of life.

with health-giving properties, certain of which supply major insights for modern medicine. It was from Amazonian tribesmen who used curare, a muscle relaxant, on their arrow tips as a hunting poison, that Western-world doctors learned of the substance's promise for surgical operations.

Other plant materials help with other medicinal problems of humankind, even anti-cancer drugs! In 1960, a child with leukemia enjoyed only one chance in five of remission. Then an American pharmaceutical corporation, Eli Lilly of

167

Top left: Australian aborigines preparing for a corroboree at Kununura. The pursuit of uranium to fuel nuclear power stations has caused considerable turmoil to the aboriginal tribes under whose land uranium ore lies.

Top right: Kalahari bushmen, whose knowledge of how to survive in the desert is unparalleled, have been badly affected by the festering war in Namibia. There are only 50,000 left.

Above: the Govoka-Asaro mudmen of New Guinea.

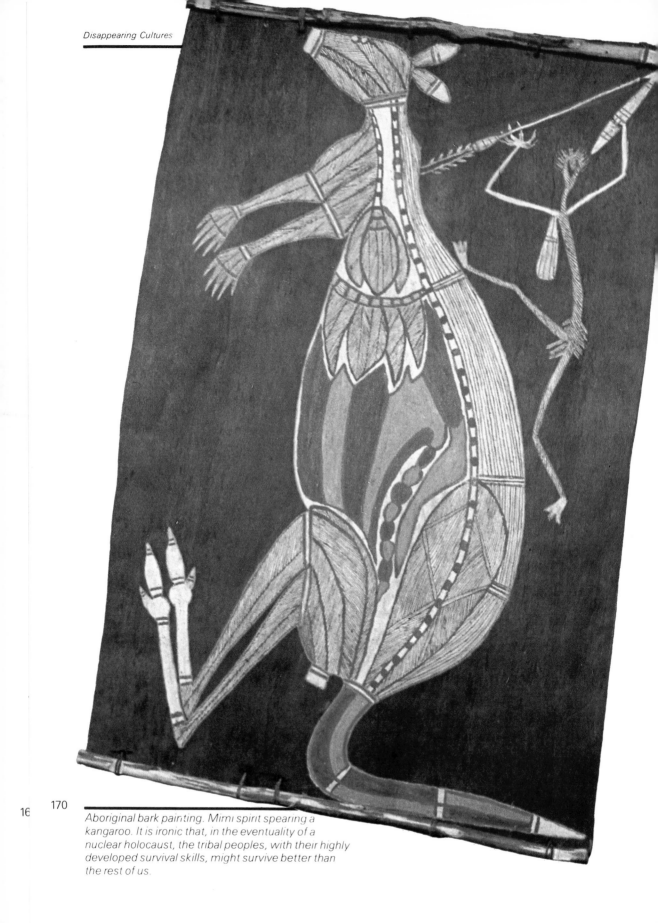

Aboriginal bark painting. Mimi spirit spearing a
kangaroo. It is ironic that, in the eventuality of a
nuclear holocaust, the tribal peoples, with their highly
developed survival skills, might survive better than
the rest of us.

Chapter 10

Will the Climate Change?

Weather changes day by day, but climate remains constant for ever. Well, almost for ever. Between one climate phase and another, with temperature and moisture changes of sufficient impact to be noticeable by humans, there can be gaps of thousands of years.

So runs the conventional wisdom. Yet this should, as Hamlet might enjoin us, be scanned. Weather appears quirky, rightly enough. We have 'big freeze' winters, like those of 1962–1963 and 1981–1982. Russian farmers will tell you that they can almost expect disastrous droughts one year followed by near-ideal growing conditions the next year, leaving them with massive food shortfalls followed by bumper harvests. Yet insofar as these phenomena occur in relative isolation, they remain transient 'blips' against the background trends. They do not establish a consistent weather pattern, enabling them to qualify as 'climate'.

In any case, climate can change pretty rapidly. Roughly between 1920 and 1960, we enjoyed a more or less benign, and above all predictable, climate – splendid news for the farmer. From about 1960 onwards, however, just when the population explosion was really starting to make its impact felt on agriculture around the world, things changed. The former steadfast patterns altered, and climatic stability seemed to disappear. Yet many

Storm clouds over Albuquerque, New Mexico.

observers would say that it is not the present phase that is out of kilter. It was the 1920–1960 phase that was the exception, and we are now reverting to a norm of up-and-down climate.

Indeed it is not going too far to say that, if our climatic crystal balls tell us anything, we should be entering a disturbed period of a scale to match the period at the start of the nineteenth century, when people suffered droughts and floods as a regular course of events, and when super-cooled winters were followed by blazing summers – all of which presented severe problems for agriculture. So far as scientists can tell, we can expect to suffer a similar period of unstable climate until the early part of the next century at least.

Another Ice Age?

Taking a longer perspective, we can anticipate even more pronounced changes. The last Ice Age ended about 10,000 years ago. Immediately thereafter came a warmer epoch, reaching a so-called optimum between 7,000 and 5,000 years ago. There then followed a period of colder conditions, with their extreme somewhere between 3,000 and 2,300 years ago. By the time the Middle Ages arrived, there was a slight warming, reaching a peak somewhere between the years AD 1000 and AD 1200. Finally came a chillier phase, popularly known as the Little Ice Age, starting around AD 1430 and enduring until about AD 1860.

These are some of the longer-duration fluctuations in climate. Within a yet more protracted time horizon, we can even anticipate that we are experiencing an overall cooling. According to some readings of the historic tea-leaves, we are due for another Ice Age proper; in fact, we may well be overdue. But we shall not examine this prospect in detail here, because the 'natural' climatic trends are starting to be over-ridden by a more dominant type of climate, a man-made climate. By the end of this century at the latest, we shall probably be witnessing

Glacier, Greenland. The glaciers, enormous rivers of ice, are remnants of the last ice age, 10,000 years ago. Now in retreat, during their advance they gouged scars on the Earth's surface which remain visible today.

some anthropogenic changes of a scale it is difficult to grasp, in that they may dislocate the very basis of our daily lives. With every passing day, evidence is accumulating to the effect that Man is modifying the conditions of his One-Earth home in fundamental fashion. For all that he is doing it unthinkingly, he is doing it effectively and increasingly.

We are conducting climatic experiments with the planetary ecosystem of a scope that we can scarcely envisage. By the time we are able to discern the shape of these changes, it will almost certainly be too late for us to slow down, let alone halt, our experiments. We are affecting the heat regimes, convection currents, wind systems and precipitation patterns of our planet in a way that (when we consider all these 'imposed perturbations' together) looks likely to cause greater environmental upheavals than anything we can imagine short of nuclear war.

Enormous Waterworks

Some of the climatic changes we may well trigger will probably be caused by our tinkering about with major bodies of water. These interventions will lead to *indirect* repercussions for climate, by contrast with a number of *direct* consequences to be described below. Yet indirect as they may be, and localized as they will surely be, they will be far from trifling. Much the best thing we can say about them is that they have not yet been attempted – whereas the other, larger effects, described later on in this chapter, are already under way.

By way of illustration, let us look at Russian plans to reverse the flow of some of their great rivers. Several great rivers flow northwards into the Arctic Sea. Were the Russians to use nuclear devices to blast new channels for them through the mountain ranges of Siberia, the rivers could be persuaded to flow southwards. The Russians are interested in the prospect for two reasons. First of all, the damp territories of western Siberia would

become dryer. Secondly, and more importantly, the Russians could irrigate new farmlands to help solve their agricultural difficulties. But set against these land-use benefits must be the climatic costs that could ensue from a diminution of the ice cover of the Arctic Ocean – with an ecological backlash that would not be confined to the northern Soviet Union, but could impinge on global weather mechanisms.

In brief, difficulties would arise if the fresh water that now flows into the Arctic along vast rivers were to be reduced in quantity. This would cause a shift, albeit a small one, in the salt content of the Arctic Ocean. The surface waters of the Ocean are much less salty than the waters several hundred meters down. Although fresh water freezes at a higher temperature than salt water, the difference in salinity between the topmost and the lower layers of the Arctic Ocean leads to a 'thermal inversion', preventing the surface ice from being melted by uprising warm waters

from the depths. The fact that the surface layers feature low salinity is due to the supplies of fresh water flowing into the Ocean from the large rivers of Siberia. With every second, about 85,000 cubic meters of fresh water are discharged by these rivers into the Arctic. Were this flow to be halted, the surface of the Arctic Ocean would no longer remain cold by grace of the salinity layering; and within just a few years the climatic regime could be disrupted. According to calculations by Dr John Gribbin, a British physicist, we may suppose that if both the Ob and Yenisei Rivers were to be diverted southwards, a sector of the Artic Ocean measuring about one million square kilometers, that is, an area about the size of Great Britain, France and West Germany together, would be immediately affected, with warmer waters near the surface of the sea, resulting in much less ice. These circumstances could lead to a feedback effect, stimulating the melting of ice over a much wider area. In turn

The Yenisei River, Siberia, USSR. One of several rivers currently being considered for reversal by Soviet planners.

RUSSIAN RIVER REVERSAL SCHEME

Possible route of main canal
Possible route of secondary scheme 300 miles (480km)

Arctic Ocean

Yenisei

330 BILLION TONNES PER ANNUM

Ob

Belogorye Ob

Tobolsk

Irtysh

25 BILLION TONNES DIVERTED

DIVERSION WORKS

PLANNED RESERVOIR

Turgai

Ural

Syrdarya

Aral Sea

UZBEKISTAN

Caspian Sea

Amudarya

again, this would influence the wind and rainfall systems extending from north-western Europe across into Siberia. In addition, the warm waters appearing on the surface of the sea would increase the amount of water vapour in the atmosphere, which could translate into greater snowfalls in cooler territories at high latitudes – which in turn yet again would affect the general circulation patterns and convection currents much further afield.

How far the ultimate effects would extend is something we can hardly speculate about intelligently, given our lack of understanding of what makes the climate tick. But we know that the Arctic tends to serve as a great 'weather machine'. For example, many scientists believe that ice cover at high latitudes is somehow linked, possibly closely linked,

with precipitation patterns in the tropics – both with too much and too little precipitation. It is not at all impossible that the rainfall regimes of the Indian Ocean, with their critical monsoonal downpours, could depend upon what happens in the Arctic.

Of course, nobody knows for sure. Nobody even has a clear idea of what is likely to happen. But we have a fair notion of what could *possibly* happen. Until we have a much better grasp of what is almost certain to happen, and until we have the agreement of all parties affected, it is obvious sense to allow the Russian rivers to persist with their established courses to the Arctic.

Thermal Pollution
To go to the other extreme, let us now

The plan is to construct a canal 2,400 kilometers (1,500 miles) long to the Amudarya rivers. This would direct some 25 cubic kilometers of water (about seven per cent) southwards from the wet lands of Siberia to the dry lands of Kazakhstan and Uzbekistan. The scheme would cost 8,000 million roubles (about £5.0 billion). Many scientists are worried that a change of this magnitude would affect the climatic system of the whole northern hemisphere.

look at the most direct influence that we believe humankind can currently exert on climate. This is thermal pollution. To date, the waste heat that we have released around the world has made no difference to global climates, the effects being confined to a strictly local level. All in all, we release from factories and so on no more than 0.01 per cent of the amount of heat that the Earth absorbs from the Sun. But if that amount were to rise to one per cent, it could lead to an overall warming of about two degrees Centigrade. Plainly we are going to have to stoke up the furnaces of industry a good deal before we start to over-cook the situation to that extent. But if we were to continue to generate waste heat at the rate at which it has been increasing during the past few decades, we would reach such an increase within one century or little more.

Amazonia's Hydrological Cycles

Let us move on now to a man-derived dislocation of climate that is rather less direct in character, but still distinctly anthropogenic (man-made). While not so global in scope as thermal pollution, it appears a more likely prospect during the coming decades. We are talking about the hydrological cycles of Amazonia, and what could happen to climate in the great

Amazonian basin if the cycles were to be disrupted.

Amazonia, the wettest region on Earth, contains two thirds of all fresh water on the face of the planet. This water is constantly falling from the sky, percolating through the vegetation, being absorbed by plants, running away into streams and rivers, and evaporating (through the Sun's warmth) or being evapotranspired (by plants) back into the atmosphere. Despite the huge size of the Amazon River system, which discharges one fifth of all river water that makes its way into the Earth's seas and oceans, more than half of the region's moisture remains within the Amazonian ecosystem. As fast as much of the moisture falls to the ground, it is returned, through the respiratory activities

Power station cooling towers in the Ruhr, West Germany. Although the heat released from the burning of fuels has not had global repercussions on climate, it can have quite marked local effects on climate. The centres of cities are typically several degrees warmer than surrounding suburbs and countryside.

Rainstorm in the Amazon, Brazil. Rainfall in the Amazon is an almost completely closed cycle that will be irreparably broken if the forests are removed.

of all those trees and other plants, into the skies, whereupon it gathers for a fresh series of thunderstorms.

Day in, decade out, the water cycles go around and around, remaining within the bounds of Amazonia. True, some additional Amazonian rainfall derives from circulation patterns outside the region, primarily from the Atlantic. But at least half originates within the region. In western Amazonia, a full 88 per cent of water reaching the ground is falling for at least a second time, and probably a third or fourth time, from the atmosphere.

According to Dr Ernesto Salati and his Brazilian colleagues, the implications are profound. Were a substantial amount of the Amazonian rain forest to be cut down, the climate would become dryer than at present. Each time a sizable sector of forest is cut down, the remainder becomes less capable of evapotranspiring as much moisture as was circulating through the ecosystem before. All this makes for a steadily desiccating ecosystem.

At what stage will the forest start to be transformed into a different kind of forest, by virtue of the drying-out phenomenon? Has the process already begun? If so, how far has it gone? Could it still be reversed? To date, we scarcely

Cape Kloostad, Antarctica. The more reflective the Earth's surface, the less of the Sun's heat is retained (see over). Fresh snow reflects 85-90 per cent of the sunlight falling on it.

know how to formulate the correct questions, let alone how to supply the right answers.

Equally to the point, we are now learning that the climate of extensive territories in Brazil outside Amazonia depend, in part at least, upon the same hydrological cycles that are so critical to the persistence of the rain forest as we presently know it in Amazonia. In the vast *cerrado* woodland tract to the south, for example, extending for 1,000 kilometers to Brasilia, there could well be less rainfall than there is right now if Amazonia were to be grossly altered. Much the same could even apply to Brazilian territories still further south and west, including the principal agricultural sectors of the country. Fortunately, the Brazilian government is starting to take note of these climatic linkages, in their full scope (so far as they can be discerned at this early stage), and Brazilians may become less inclined to look upon their Amazonian forest ecosystem as a 'valuable asset going to waste'.

The Albedo Effect – the Mirrors of Earth

So much for the consequences of deforestation in a limited sector of the tropical forest zone. What if deforestation were to become a pervasive phenomenon throughout the tropical forest zone, in Africa and Asia as well as Latin America?

We encounter a rather different picture when we examine the 'albedo effect' – a phenomenon that refers to the proportion of sunlight that the Earth's surface reflects back into space. Let us suppose, for the sake of argument, that the Sun radiates 100 units of energy per minute in the direction of the Earth. When this energy reaches the outer limits of the atmosphere, the atmosphere may well absorb 20 units, allowing only 80 units to strike the Earth's surface. If, due to the particular vegetation pattern that occupies that part of the Earth's surface, some 40 of these 80 units of energy are reflected upward and back out into space, the albedo is reckoned at 50 per cent.

In point of fact, the amount varies a great deal. Fresh snow, with its high reflectivity, has an albedo of 85–90 per cent. Deserts, with little vegetation to absorb and retain the sunlight, generally reflect 25–30 per cent. Croplands, with their moderate amounts of vegetation, span a range of 12–20 per cent albedo. Tropical evergreen forests, with their dense masses of vegetation, reflect only 7–15 per cent. Because of its impact on convection patterns and wind currents, and hence rainfall regimes, the albedo effect constitutes a fundamental factor in controlling climate around the world.

Although we have no conclusive scientific evidence as yet, because we do not yet understand precisely all the factors involved, many scientists believe that widespread clearing of tropical forests

Coniferous forest and clearings in Germany. The different albedoes of croplands and forests are clearly shown.

would cause an increase in the 'shininess' of the Earth's surface. As a result, the deforested lands would reflect more solar heat than before. Much of the tropical forest zone, located in a broad band on either side of the equator, corresponds with a zone where there is a pronounced up-rising of air, a process that introduces energy into the vast circulatory movements of the general atmosphere – and if this process becomes disrupted through deforestation, the repercussions could well extend, through altered patterns of air circulation, to the temperate zones, especially those of North America and Eurasia (not so much in the southern hemisphere, because of the modulating effect of the ocean expanse).

So far as climatologists can tell, a not unlikely outcome would be a decrease in rainfall in the equatorial zone itself, an increase in rainfall for territories between five and 25 degrees North and South, and a decrease for lands between 40 and 85 degrees in the North. In short, greater rainfall for the southern half of the Sahara, India and much of the arid sector of Mexico, but reduced rainfall for the northern half of the United States and Canada, and for most of Europe and the Soviet Union. Plainly these climatic dislocations could prove traumatic for food-growing territories of the northern temperate zones, notably the grainlands of North America, Europe and the Soviet Union.

The Greenhouse Planet

We often think of carbon dioxide as an insignificant gas, of most direct relevance to our daily lives when it supplies the 'fizz' to soft drinks. But carbon dioxide exerts a much greater effect than that. While it constitutes only 0.03 per cent of the gases in the Earth's atmosphere (by contrast with nitrogen's 78 per cent, and oxygen's 20 per cent), it serves to absorb radiant energy from the Sun. The way in which carbon dioxide affects the temperature on Earth is by what is called the 'greenhouse effect'. It behaves rather like the glass in a greenhouse, letting the heat from the Sun through to the Earth, but preventing some of it from being radiated back out through the atmosphere. The carbon dioxide traps infrared radiation in the atmosphere, keeping the Earth at a higher temperature than it would be otherwise. In terms of the global heat balance, the carbon dioxide blanket has – until the advent of industrial society – ensured that the average temperature of the surface of the Earth is about 15°C. But now that the amount of carbon dioxide in the atmosphere is increasing, that value also seems bound to increase – unless other, unrelated effects should by chance happen to compensate for it by tending to produce lower temperatures.

The amount of carbon dioxide in the

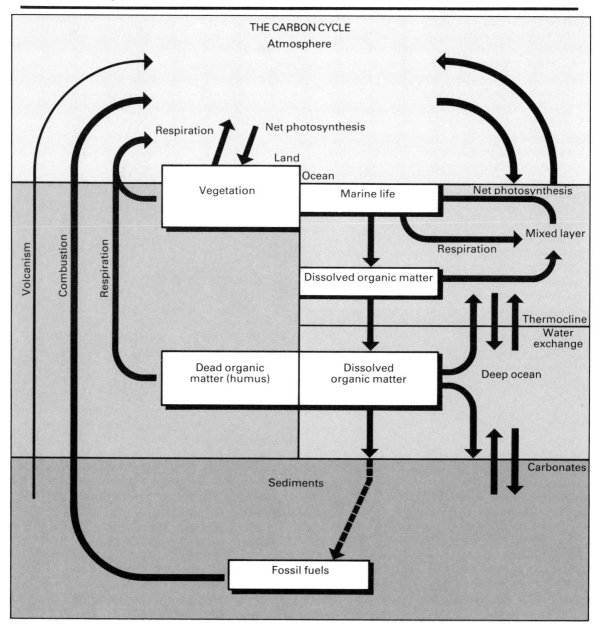

THE CARBON CYCLE
Atmosphere

Respiration   Net photosynthesis

Land
Ocean

Vegetation   Marine life   Net photosynthesis

Mixed layer

Respiration

Volcanism   Combustion   Respiration

Dissolved organic matter

Thermocline
Water
exchange

Dead organic
matter (humus)   Dissolved
organic matter   Deep ocean

Carbonates

Sediments

Fossil fuels

atmosphere has fluctuated in the past, possibly in relation to the global temperature – although this is disputed. Studies of air samples found in bubbles trapped in Antarctic ice, dated to the coldest period of the last Ice Age, show concentrations of carbon dioxide of only half the present level. When the oceans are colder, they dissolve more carbon

dioxide – which might account for changing temperatures. But those who dispute the theory point out that more absorption by oceans followed by less carbon dioxide and a weaker greenhouse effect and thus colder temperatures and more absorption would create a downward spiral from which no exit appears possible. And they point out, too,

Global flows and pools of carbon. The boxes represent 'pools' — places where carbon accumulates, the arrows show the flow between pools.

that the quantity of carbon dioxide in the atmosphere is relatively small – only about one thirtieth of one per cent.

If the build-up of carbon dioxide in the global atmosphere continues, the result will probably be warmer temperatures for the planetary ecosystem as early as the second quarter of the next century, together with basic shifts in rainfall patterns for extensive sectors of humankind's habitats. It could further lead, in the longer term, to some melting of polar ice, with a marked rise in sea levels. Overall the economic consequences, plus the 'political fallout', of this phenomenon could prove far-

additional 600 billion tonnes of carbon. The planetary atmosphere contains almost 700 billion tonnes, which now works out to about 335 parts per million – a concentration that is increasing at a rate of about 1.5 parts per year. Finally, the Earth's biotas: living things contain about 800 billion tonnes, while other sectors of the biosphere contain at least 1,000 billion tonnes (possibly much more) in the form of soil humus and peat. To give an idea of the way the biosphere's carbon stock is divided up among categories of living things, plants are believed to contain about 800 billion tonnes – of which 80–90 per cent is in forests, almost

reaching indeed.

The Earth's carbon cycle involves a stock of roughly 40,000 billion tonnes. Of this stock, the great bulk, some 35,000–38,000 billion tonnes, occurs in the form of dissolved inorganic carbon in the deep oceans. The oceans also contain a further 3,000 billion tonnes of dissolved organic matter, while the surface layers contain an

half of it in tropical forests. Humankind, by contrast, with over 4.6 billion individuals averaging less than 50 kg in weight and a carbon content of about 4 kg each, comprises well under 20 million tonnes.

During the course of natural processes, the carbon stocks that contribute to the global cycle are slowly

Icebergs, Weddell Sea, Antarctica.

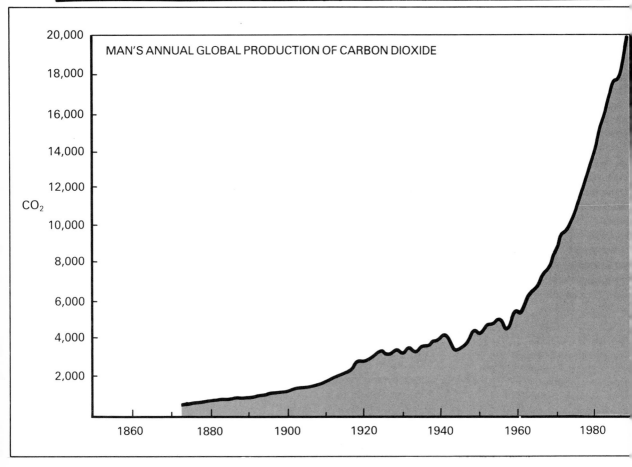

MAN'S ANNUAL GLOBAL PRODUCTION OF CARBON DIOXIDE

turning over. Each year, plants exchange more than 100 billion tonnes of carbon with the atmosphere (and the oceans almost as much); and every 300 years or so, plants, in conjunction with other living things, are believed to cycle all atmospheric carbon dioxide (by contrast with oxygen, every 2,000 years, and water, every 2 million years). In the carbon cycling process, the oceans plainly play a major role, in that they serve as giant reservoirs for carbon: of the 180 billion tonnes of carbon dioxide added to the global atmosphere between 1850 and 1950, about half has been transferred into the oceans.

At roughly the midpoint of the last century, when the Industrial Revolution was gathering momentum, the atmospheric concentration of carbon dioxide was about 290 parts per million, possibly a little less. As industry stepped up the burning of fossil fuels, the concentration increased. By 1960 it had risen to 310 parts per million, and the past 20 years have seen an increase of well above a further 20 parts per million. At the current rate of increase, which is itself increasing, the pre-1850 'natural' amount could well double by the year 2030 or soon thereafter.

Where Does the Carbon Dioxide Come From?
The principal culprits believed to be guilty are the factory and the automobile. These two between them are generally believed to be contributing the largest amounts of

Man's annual production of carbon dioxide from burning fossil fuels and from cement works rose steadily from the middle of the 19th century and more steeply since 1940. (These figures, in million tonnes, do not include CO_2 from burned forests).

carbon dioxide injected by humankind into the skies each year. Since the benchmark date of roughly 1850, at least 150 billion tonnes of carbon have been released into the atmosphere through fossil-fuel combustion. Of this amount, only half is thought to have remained in the atmosphere, the rest being absorbed by the oceans, and by other, hitherto unidentified, large 'sinks' for uptake of carbon dioxide. During the period 1960–1980 alone, some 80 billion tonnes of carbon are thought to have been emitted into the atmosphere, roughly half of it being absorbed by sinks.

Right now, we are burning fossil fuels at a rate of around five billion tonnes a year. Moreover, we are increasing our burning rate at a little over 4 per cent per year, and if we continue with this level of growth (albeit there has been a slight slackening since OPEC set up shop), we could eventually cause the atmospheric concentration of carbon dioxide to rise to 600 parts per million. If we slow the growth rate of fossil-fuel combustion to only 2 per cent per year, the doubling time will be extended by no more than 15–20 years. If we were to throttle back on our increasing use of fossil fuel, and maintain today's level into the indefinite future, the doubling would be delayed until well into the twenty-second century.

In all this picture, there is a role played by tropical forests. Insofar as much deforestation is accomplished through burning of trees to make way for agriculture, rather than through felling of hardwood timber for parquet floors and fancy furniture, many observers believe that the destruction of tropical forests is a significant source of atmospheric carbon dioxide. When a tract of forest is burned, not only is carbon from living plants released, but further carbon becomes available when the soil is exposed to the tropical sun, and soil humus with its organic carbon becomes oxidized. It is true that some forest cover that is disrupted by small-scale farmers grows back again, to make up a forest formation

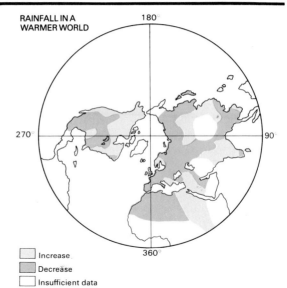

RAINFALL IN A WARMER WORLD

☐ Increase
▨ Decrease
☐ Insufficient data

with fewer species but with greater capacity to produce large amounts of plant biomass in short order – and to this extent, a successional restoration of forest cover will counterbalance some of the depletion activities. On the whole, however, there is much evidence to the effect that forest farmers generally serve to reduce the amount of forest biomass. That is, their net impact amounts to an increase of the carbon dioxide in the global atmosphere.

How much carbon may we suppose is released from tropical forests? We are far from reaching any conclusive answers. But for the purposes of this analysis, let us consider that a reasonable estimate, proposed here strictly as a 'working appraisal', could be that the net amount of tropical forests eliminated through burning each year amounts to perhaps one per cent or so of the biome. This translates into an amount of somewhere between two and four billion tonnes of carbon – or between two fifths and four fifths as much carbon as is released through combustion of fossil fuels. Furthermore, we can anticipate that as forest farmers become much more numerous in the next few decades, they are likely to step up their rate of burning.

Analysis of rainfall patterns in warm years since 1925 suggests what a warmer climate might be like. The US and USSR grain belts get less rain, while India, the Middle East, parts of Africa and most of China get more. In terms of food production a global warming would not benefit the two superpowers, but might help large parts of the Third World.

It is far from improbable that tropical forest burning will eventually catch up with fossil-fuel combustion as a source of carbon dioxide in the global atmosphere.

A Warmer World

So much for some 'educated assessments' of what is happening to the global carbon budget. Let us now move on to another area of analysis, where the situation is beset with still more uncertainties – and where the outcome, in terms of upheavals for humankind's welfare, could be significant to an extreme degree. We have already noted that the greenhouse effect implies that a doubling of atmospheric carbon dioxide beyond the baseline figure of 290 parts per million in the middle of the last century could lead to an increase in global temperatures of a mean 3°C, causing temperatures to rise toward a level not known on the planet since the Mesozoic, the age of the dinosaurs. In point of fact, there would be an increase of only 1°C or little more at the equator, around 3°C in temperate latitudes, and as much as 4.5°, possibly up to 7°C, at the poles.

Temperature changes of this order will almost certainly be accompanied by shifts in rainfall patterns. There could be an overall increase in rainfall of about 7 per cent. But – and herein lies a key factor – there would be a good deal of differentiated distribution. At present the main rainfall zones are as follows. First an equatorial zone of heavy rainfall; then a sharp switch to the anticyclone zone with its low rainfall (these two belts either side of the equator feature almost all the world's hot deserts); next a zone of prevailing westerly winds and temperate moist climate; and finally the polar regions with their markedly reduced precipitation. Were a 'greenhouse effect' to come into play, these climatic zones would tend to shift polewards, more pronouncedly in the northern hemisphere (with its larger land mass) than in the southern hemisphere. Certain zones that are now dry, such as the Sahel along the southern fringe of the

Sahara, could receive the benefit of equatorial rain, but the Sahara itself could extend further north, with adverse repercussions for conventional agriculture in parts of the Mediterranean basin – as in the extensive food-growing territories of the Soviet Union where drought is already a pervasive threat. Still further northwards, a warmer climate could extend the presently short growing season for grain crops in both North America and Eurasia (each increase in average temperature of 1°C can lengthen the growing season by about two weeks). But whereas there are good soils available for the Russians in Siberia, and we may eventually witness wheat fields on every side in Scandinavia, even Iceland, North America's grain belt cannot readily migrate northwards, since it will encounter thin and infertile soils. A temperature increase of just 1°C could reduce the US corn crop by at least 11 per cent, especially when linked with greater evaporation and a decline in August rainfall; while the same climatic dislocations could depress the wheat-growing industry by at least half a billion dollars (1980 prices) per year. In short, the great American grain belt could start

Russian wheat harvest, Stavvopol Territory. By affecting food production, climatic changes could alter the balance of power between superpowers.

to come unbuckled.

Let us gain an insight into the way these climatic scenarios might work out, by looking at precipitation patterns during the 'climatic optimum' of 6,000 years ago. At that stage, there appears to have been wetter conditions in much of Europe, the Middle East, China, India, and parts of Australia, whereas (just as is predicted by the carbon dioxide model) much of North America appears to have been drier. Were the rainfall map in the middle of the next century to resemble that of 6,000 years ago, many countries could benefit, while many would suffer. India might find itself with more rain, while neighbouring Pakistan and Bangladesh could find themselves with less: the political implications could be so extensive that it is difficult to visualize the power upheavals. Similarly, Mexico might turn into a bread basket, thus adding much to that nation's politico–economic clout. Several Islamic countries of northern and north-eastern Africa, as of the Middle East, could find themselves with a 'good weapon' to restore their international muscle after their oil wells run dry. Perhaps most significant of all, the Soviet Union and China might find themselves with such enhanced capacity to grow food that they could hope to dictate terms to the rest of the world on a whole range of issues.

True, these scenarios remain little more than speculation at present. We simply do not have enough information about how climatic changes are likely to work out, let alone what the agricultural 'backlash' will be. A carbon dioxide build-up could have all manner of unanticipated positive impacts. For example, it could exert a fertilizing effect on crops by stimulating photosynthesis, thereby enabling certain areas with reduced rainfall to nevertheless maintain their agricultural productivity. Certain agronomists calculate that if total plant biomass were to increase by only about three quarters of one per cent, that would take up all the carbon from fossil fuels, five billion tonnes, added to the atmosphere each year, and probably only half as much again could well account for all carbon emitted from tropical forests.

Much remains unknown. What we do already know, however, is that during the past 100 years, global temperatures have increased by about 0.4 C, half of the increase having occurred during the short span from the mid-1960s to 1980. These temperature increases are consistent with the calculated greenhouse effect due to measured increases of carbon dioxide.

The Sinking Cities
Significant as such changes in precipitation patterns would be for agriculture, there could emerge a set of even more serious environmental repercussions if global temperatures were to rise to a level where they could trigger a melting of the polar ice packs. A mean increase of much above 5 C would probably cause the ice pack of the Arctic Ocean to disappear completely during the summer, although parts of it may reappear in winter. An ice-free Arctic Ocean is a phenomenon that has not occurred at any stage during the past one million years.

Similarly the Greenland ice cap could start to melt, although the process would be slow and protracted. By contrast, the West Antarctic ice sheet, lying on bedrock below sea level, could break away from its 'moorings' and slide into the sea, whereupon it could break up rapidly. By the second quarter of the next century, this huge Antarctic ice pack could be disintegrating, and it could well melt away within as little as 100 years.

These changes in the polar ice packs would cause sea levels to rise by between five and seven meters before the end of the next century. In the longer-term future, during a period of several centuries, both polar ice packs could conceivably be eliminated altogether, which would raise sea levels by 50–70 meters.

Suppose sea levels were increased by five to seven meters. Many low-lying land areas, such as Florida and the Netherlands, would lose much of their territory beneath the waves. In fact, at least 40 per cent of Florida's population would be affected, and more than eleven million people in the United States as a whole. Presumably the flooded-out people could be settled elsewhere, albeit at massive cost, and supposing there were plenty of time for planning. Similar problems would arise for coastal cities and other concentrations of human communities in other parts of the world, involving two fifths of all humankind: more than 30 per cent of all people live within a 50-km zone adjoining seas and oceans. The American economy could presumably afford to 'move New York inland', and the same for Los Angeles, New Orleans and other coastal conurbations. A similar process would have to be implemented with such other cities as London, Glasgow, Tokyo, Osaka, Montreal, Stockholm and Copenhagen. But people in Calcutta, Shanghai, Jakarta, Cairo, Lagos and Rio de Janeiro would have less room to manoeuvre, since their hinterlands are already crowded to bursting point, and the national economies in question would have far

ANTARCTIC ICE AND THE EASTERN USA

8m 26ft
30m 98ft

fewer funds to cater for mass migrations.

What should we do to respond to these various scenarios? Obviously we can take steps to damp down our appetite for fossil-fuel energy. This is in fact occurring in North America and Western Europe, which now experience growth rates down to somewhere between zero and 2 per cent per year. But Eastern Europe and the Soviet Union (which could prove to be beneficiaries from a carbon dioxide build-up) are expanding their fossil-fuel consumption at 4 per cent a year. As for the developing nations – which might construe a call by the developed nations for a cut-back on use of fossil fuels as a further ploy of the developed world's 'conspiracy' to block their progress toward industrialization – they are stepping up their appetites by 5 per cent a year. All

This map shows areas submerged in the interglacial past when the sea was 8 metres (26 feet), and 30 metres (98 feet) higher than today. If the West Antarctic ice sheet melted, the ocean might rise 5 metres (16 feet) in a few decades.

this means that North America, which now accounts for almost 30 per cent of all commercial energy used, may well account, by the year 2025, for only 10 per cent, while the developing nations could increase their share from 13 per cent to 40–50 per cent.

Planting Trees to Soak Up CO$_2$

The day may come when we find that a major response to the carbon dioxide problem lies, quite simply, with planting trees. Man-made forests can soak up immense amounts of excess carbon dioxide from the atmosphere. Present tree plantations amount to well below one fifth of what reflects immediate needs for timber, let alone future efforts to safeguard climatic stability. Ultimately, however, there may be much advantage for us, in terms of climatic stabilization, in the establishment of extensive plantations in the humid tropics where growth of biomass is fast and continuous. But how many trees, roughly speaking, would we need to plant?

Well, we know that the present increase in atmospheric carbon dioxide each year amounts to about 2.5 billion tonnes. It is realistic to project that, if the burning of fossil fuels and of tropical forests continues to expand at recent rates, the quantity could increase to four times as much by the year 2000 or shortly thereafter. For present purposes, let us work with a figure of five billion tonnes a year. Were we to embark on a gigantic 'soak up' campaign, we could use, for the sake of illustration, a tree such as the American sycamore, which now grows faster than most other species in the plantations of Georgia in the United States. Wood of course is not just an aggregation of carbon. It is a complex mixture of carbon and hydrogen, together with some oxygen and nitrogen. A healthy growing sycamore can add as much as 45 kg of wood to itself each year. Since we need to think in terms of growing a volume of wood of some 10^{10} (ten billion) tonnes a year, we would have to

envisage a reafforestation effort that would require about 5×10^{11} (500 billion) trees to do the job. To put the point another way, we could hope to transfer carbon from the atmosphere and into tree tissue at a net rate of about 7.5 tonnes per hectare per year; so to soak up the excess carbon dioxide, we would need to think in terms of a total area of seven million square kilometers, or a tract almost as great as the area of the United States (minus Alaska and Hawaii).

Of course, these figures are presented for ease of ready reckoning, even though they refer to a temperate-zone tree growing in a sub-tropical zone, that is, under less than the optimal conditions we could expect to find in the humid tropics. The better plantations of eucalyptus, ipilipil and pine in the humid tropics can grow new wood more than twice as fast as does the sycamore in Georgia. So if the effort were confined to humid tropics (supposing sufficient space could be found), we could perhaps make do with an area of only 3.5 million square kilometers, or an area roughly equivalent to Brazilian Amazonia. The famous Jari enterprise, near the mouth of the Amazon River, amounts to only 15,000 square kilometers, a little larger than Connecticut. All tree plantations in the humid tropics now comprise about 100,000 square kilometers. So we would have to think in terms of a sizeable patch of real estate should we ever seek to resolve the carbon dioxide problem by growing trees.

We would also have to think of the enormous costs for the necessary fertilizer. Tree plantations that produce large amounts of wood year after year generally require regular doses of fertilizer. Drawing on experience with the American sycamore again, a super-scale reafforestation effort would be likely to consume fertilizer annually equal to half the amount already being used around the world, mainly for agriculture. This would cost around twelve billion dollars (roughly half the cost of sending a man to the Moon). In addition, there would be the

expense of establishing plantations,
around 600 dollars per hectare as a
minimum. This all suggests a total sum
way above 200 billion dollars. Yet large as
these amounts sound, they are trifling
when compared to the costs that would
arise in the wake of a melting of the polar
ice caps, which are tentatively estimated
at a minimum of one thousand billion
dollars.

Norman Myers

Reafforestation in the Amazon.

Chapter 11

The Politics of Ecology

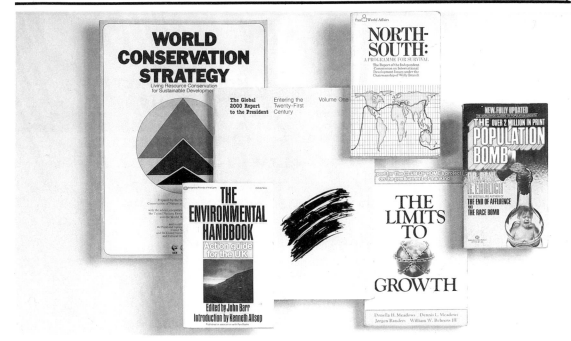

Before 1972, there were no real politics of the environment. Until the United Nations Conference on the Human Environment which took place in Stockholm in May and June of that year, concern for the fate of the environment had been the private preoccupation of a small number of scientists, administrators and enthusiasts involved in nature conservation. Words like 'ecology' and 'pollution' had only just begun to invade our daily newspapers.

For the great mass of people, Nature remained what it had always been, a vast unknowable force: sometimes hostile, sometimes bountiful, often beautiful, occasionally terrifying, but always present, always larger than Man. There were no departments of the environment, no environment correspondents, no ecological political parties. Schools and universities did not offer courses in environmental studies. Until the 1960s, no one thought that we might run out of essential natural resources and the idea that human decisions might unravel the intricate and resilient web of life on which all living beings depend seemed too ludicrous to consider.

Yet, only a few years later, all this has changed with a suddenness and completeness that makes it almost impossible to remember a time when a fear for the future of the planet was not with us. In the most extensive and swiftest public education effort in history, hundreds of millions of people have been alerted to the growing threats to their survival. There are now environment and natural resource management agencies in 144 countries. Most serious newspapers have an environment correspondent. The range of environmental education

The books that alerted the world.

opportunities is bewildering, and in Western Europe there are more than a dozen ecological parties seeking votes.

The pressures that generated this explosive expansion of human awareness had been building for a long time from the early part of the nineteenth century onwards. As the environmental effects of the industrial revolution became apparent, scientists and private individuals had begun to build the organizations which are the foundations of the present environment movement. The drive for prosperity in the aftermath of World War Two led to a massive expansion in world trade and a rate of technological development that multiplied man's harmful impact on the environment many times over. Inevitably, this produced a further response from concerned people.

By the late 1960s, incidents like the Torrey Canyon oil spill and books like Rachel Carson's *Silent Spring* (1962 – it was the first popular ecology book, a bestseller) had alerted public and mass media alike to the rising tide of pollution. Other publications, such as Paul Ehrlich's *The Population Bomb* (1968) and the Club of Rome's *Limits to Growth* (1972) had raised the possibility that we might run out of vital natural resources. New organizations such as Friends of the Earth and the Conservation Society were founded to fight pollution and the wasteful use of resources.

The 1960s were also a time of much soul-searching, in the more affluent nations at least, about the increasing dominance of materialist values. Growing numbers of people began to feel that the pursuit of more and more possessions was insufficient as a purpose for life. Surrounded by the mass-consumption, throwaway society, many people – especially among the young – began to search for a simpler and more satisfying way of life in which they would be more in touch with each other and with nature.

It was these rivers of opinion on nature conservation, on pollution, on resource use and on new lifestyles running together which produced the flood of public alarm that led to the Stockholm Conference. What that Conference marked was not just the recognition that the environment was facing serious problems, but also that solving these problems was a *political* question that could not be left to the experts and to private individuals.

Not only was it recognized that governments must act if mankind were to avoid destroying the basis of life, but it was also recognized that they must act together since environmental problems were no respecters of national boundaries. The result of the Stockholm Conference was thus to establish, for the first time, the place of the environment on the political agenda.

The Glasgow Company's works at St Rollox, 1859..

The hippy culture of the 1960s was inspired in part by a desire to re-establish harmony with Nature.

Prior to the present, the absence of the environment from the political agenda was unremarkable. Human decisions, however important to human beings, had, at worst, only local effects on the destiny of other life on the planet. What has changed during the last century, but especially since 1945, has been the huge increase in man's ability to transform the environment. The conquest of nature, an ancient goal of our civilization, has been substantially achieved. The benefits to the health and longevity of our population, to the safety and comfort of our dwellings and for the quantity and reliability of our harvests, have been welcome. The responsibilities, above all to use our new-found powers wisely, have been less readily accepted.

We have unlocked the secrets of the atom, as was so dreadfully demonstrated at Hiroshima. The knowledge thus gained has transformed our lives. But the fragility of the political mechanisms, both national and international, that hold such power in check is all too evident in the growing arsenals of nuclear weapons. Nothing could be more catastrophic for the environment than an all-out nuclear war, yet as the number of nations with the knowledge and means to make nuclear weapons grows, so too do the chances of such a war.

We are now busily unlocking the secrets of the living cell and in so doing not just acquiring the ability to improve agriculture and medicine, but also to manipulate the whole course of evolution. Our power to synthesize wholly new chemical compounds is introducing new poisons into the environment each year, often with little knowledge of their effects on the health of human beings and natural systems. Our enhanced ability to clear forests, build roads and dams, drain wetlands and excavate minerals is destroying natural habitats on a scale never before experienced by the planet.

Yet none of our political institutions, rooted as they are in the nineteenth century, was designed to contain such capabilities. The radioactive wastes produced by today's nuclear power stations will have to be safely managed for periods longer than the whole of recorded human history. Intensive agriculture produces high incomes for farmers now at the expense of less productive land for future generations. Trees cut from steep hill slopes in one country cause floods in

A disrupted frog virus releasing its DNA molecule (which carries its genetic coding).

A story for our times.

another, toxic wastes emitted from power stations in Britain fall as acid rain, destroying trees and fish, in Sweden. The politics of ecology are the politics of interdependence, not just between nations, but also between generations.

There has rarely been an era when the ancient myth of the Fall was more relevant. In a very real way, mankind has 'eaten the fruits of the Tree of Knowledge of Good and Evil' in the last 40 years. We no longer live in a world in which care for the environment as a whole is none of our concern, since we have acquired – for the first time ever – the knowledge and power to destroy the whole of that environment. Responsibility for the future of all life on the planet has now fallen on each one of us. The fate of the Earth will not only be decided by the impersonal forces of Nature but also by our choices, expressed through the action, and inaction, of the political process. Thus the politics of ecology is also the politics of survival.

Although the Stockholm Conference placed the environment on the political agenda, it did not place it very high up. Governments do not allocate budgets for environmental action. Policies have been developed and agencies established to implement these policies. But placing an issue on the political agenda does no more than bring it into competition with those issues already there. A central task of eco-politics is to resolve the conflicts with other political priorities – reducing unemployment, raising living standards, curbing inflation, providing welfare services – in such a way that the basic sustainability of biological systems is not undermined. Just how complex this task can be will become apparent as we explore in a little more detail the politics of just three of the many environmental problems.

The Politics of Population
Sir Peter Scott is one of the most famous figures in the global environment movement. Few people have done more than he to move public and political

opinion on the future of our ecology. In June 1982, he addressed an invited audience of equally distinguished people from around the world at a meeting to commemorate the tenth anniversary of the Stockholm Conference. There was, he told the assembled guests, one person who could do more than all of them together to advance the cause of the environment – the Pope.

In launching this understated, but effective, attack on the repeated reaffirmations by Pope John Paul II of existing Roman Catholic dogma on birth control and abortion, Sir Peter was reopening an issue that had riven the original Stockholm Conference. On that occasion, with the shock waves from Paul Ehrlich's *The Population Bomb* still reverberating around the world, there had been a vigorous, occasionally virulent, debate on the relationships between population and declining environmental quality. For some there was, in Gandhi's famous words, 'enough for man's needs, but not for his greed'. The problem was not to limit population size so much as to ensure that the available resources were

Top: Pope John Paul II, Latin America, 1982.

distributed more fairly. Their opponents argued that it would simply not be possible to sustain a rapidly growing global population at the levels of affluence enjoyed by the industrialized countries no matter how equitably resources were distributed.

To some people from the South this latter argument sounded like another defence of Northern privilege: nothing more than a blind to disguise an unwillingness to share resources. From their perspective, each Northerner in using up many times the resources used by a Southerner, was equivalent to many more than one person in his harmful impact on the environment. The real overpopulation problem was that of greedy Northerners, not poor Southerners. To many in the North these arguments seemed like an evasion of responsibility. Northern populations were stable, or even

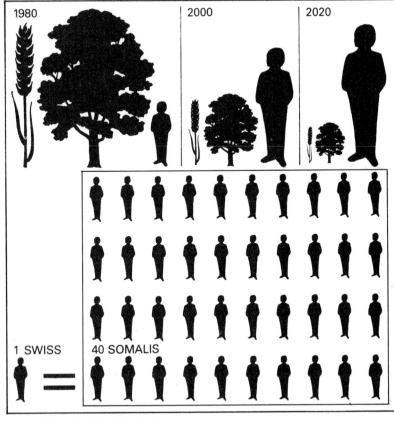

POPULATION AND RESOURCES

Living resources essential for human survival and sustainable development are increasingly being destroyed or depleted. At the same time human demand for those resources is growing fast. If current rates of land degradation continue, close to one third of the world's arable land (symbolized by the stalk of grain) will be destroyed in the next 20 years. Similarly, by the end of this century (at present rates of clearance), the remaining area of unlogged productive tropical forest will be halved. During this period the world population is expected to increase by almost half. The predicament caused by growing numbers of people demanding scarcer resources is exacerbated by the disproportionately high consumption rates of developed countries.

Source:
World Conservation Strategy

1980 2000 2020

1 SWISS 40 SOMALIS

beginning to fall, and all the efforts of the North to help Southern countries develop were being negated by their own failure to check their rapidly rising populations.

As is so often the case in disputes of this kind, there was a kernel of truth in both positions. It has since become much clearer that poverty itself is a cause of overpopulation. (*See also* Chapter Three.) Poor people are unaware of the means of controlling pregnancies and are often too poorly educated to use those means even if they were available. Poor countries

money intended to aid economic development into Swiss bank accounts and luxury consumption for the ruling elite. Much-needed capital is spent on unnecessary weapons and the soldiers to use them, or on prestige projects like dams and airports which rarely bring much useful benefit to the mass of rural poor. Personal greed rather than public need usually concentrates national investment in the cities (*See* Chapter Eight) rather than in the rural areas where poverty is most marked.

cannot afford systematic family planning campaigns. Poor families see more children as an insurance against high rates of child mortality, and as a provision for their old age. Thus, to the extent that Northern countries are unwilling to make the changes in the world economy necessary to banish poverty from the South, they are themselves partly to blame for the overpopulation of which they fear the consequences.

But it is also true that many developing countries (often called the 'Southern' countries because that is where most of them are), with the most notable exception being China, have made little serious effort to control their burgeoning populations. Corruption often diverts

All our experience suggests that what slows population growth down is a combination of economic progress and a vigorous population policy. Economic progress brings with it education, health and security, all of which encourage smaller families. An active population policy provides the means to translate this wish into falling birth rates and a social climate which rewards smaller families. When this combination occurs, as it has for instance in Taiwan and South Korea, population growth rates begin to fall quite fast.

In attempting to understand the political implications of the population problem it is often difficult to remember just how large the problem is. World

In six hours the world spends as much on arms as it has spent in ten years in supporting the United Nations Environment Programme. Seventy-five per cent of the trade in conventional weapons consists of sales of arms from the industrialized countries to the Third World.

population is increasing at a rate of more than 1,000,000 people every five days. Between now and the end of the century we will add about two billion people to the population. Simply to house this number of people would mean building a city the size of London every twenty days from now until the year 2000. But just housing them will not be enough. They will need to be fed and found employment, provided with education and health services and supplied with fuel and water. In a world already beset by

gross inequalities, faced with persistent recession and continued international conflicts, finding a place to consider the needs of two billion additional passengers on Spaceship Earth will not be easy.

It is not as if we do not know what to do, or how to do it. Stabilizing world population at about six and a half billion people would at least leave us a chance of achieving a sustainable environment. It would mean the industrial countries achieving a stable population by the year 2000, a goal many of them have already achieved. It would mean developing countries reducing their birth rate from the current annual rate of 3.2 per cent (32 new babies a year for every thousand existing people) to 2.6 per cent by 1990

and to about one per cent by 2020. This rate of reduction is no greater than that already achieved by a number of countries such as China, Indonesia and Costa Rica. The means to implement this strategy by economic progress and family planning were discussed above. What seems to be missing is the political will to implement these means.

Recent calculations suggest we should be spending about three billion dollars a year on the family planning side of the equation, a sum so small as to be

inconsequential when compared with global military expenditure, which has been six thousand billion dollars in the past ten years. Instead we are spending only one third of this amount. Global economic growth seems likely to average closer to 2 per cent than the 4 per cent of recent years. This will leave half the world's population living in countries in which population growth exceeds economic growth with obvious consequences for economic progress. As these countries struggle to survive, and as they eat into their environmental capital to do so, so the spiral of poverty leading to population growth, leading to environmental degradation, leading to more poverty will take another downward

Dinka mother with twins, Sudan. Less than a half of one per cent of our expenditure on arms would pay for a population stabilization programme.

Mercedes in a Nigerian village. Inequality is an obstacle to protecting the ecology.

twist. The world will become even more divided into two camps, in one of which economic growth exceeds population growth and where living standards are rising, and in the other of which population growth exceeds economic growth and living standards are falling.

The global problems of the twenty-first century, those we should be beginning to solve now, will require more foresight, co-ordination, trust and mutual understanding by more people than has previously been displayed in human history. We will be living in a world in which small nations, or even terrorist groups, might have nuclear weapons; to which science will have delivered even greater chemical and biological power, to be used for evil as well as good, and which will be armed and frightened as never before. Managing this fragile world so as to sustain the biological basis for life will require great advances in the political maturity of people and their governments. Such advances will not take place in a world increasingly divided by population pressures.

The Changing Pattern of World Grain Trade*					
Region	1934-38	1948-52	1960	1970	1980
	(million metric tons)				
North America	+ 5	+ 23	+ 39	+ 56	+ 131
Latin America	+ 9	+ 1	0	+ 4	− 10
Western Europe	− 24	− 22	− 25	− 30	− 16
E. Europe and USSR	+ 5	0		0	− 46
Africa	+ 1	0	− 2	− 5	− 15
Asia	+ 2	− 6	− 17	− 37	− 63
Australia and N.Z.	+ 3	+ 3	+ 6	+ 12	+ 19

* Plus sign indicates net exports; minus sign, net imports.
Source: Food and Agriculture Organization, U.S. Department of Agriculture, and author's estimates.
Source: Worldwatch Institute

Top: we must expect that by the end of the century many more than five nations, and possibly even some sub-national groups, will have acquired nuclear weapons.

Above: these charts show the growing dependence of the world on just two regions for its supplies of grain. In the 1930s six of our seven regions were net exporters. Now there are only two exporting regions. Both Asia and the Soviet Union have moved from surplus to huge deficits, and the growing dominance of the USA is apparent.

The Politics of Food

Politics is the child of agriculture. It was only when mankind learned to drain the rich wetlands of the river deltas that politics became a part of life. Agriculture brought with it the need to define, and defend, a permanent territory. It also brought with it the ability to produce and store surpluses of food that could be transported and distributed. Politics is the process by which we decide who gets access to what. Since the invention of agriculture, deciding who will eat has been, as it remains, the most fundamental of all political questions.

One reason why the Soviet Union is unlikely to launch a nuclear attack on the United States is that North America currently supplies over 12 per cent of the grain needed to feed the Soviet people. These supplies would not be easily available in the aftermath of a nuclear war, even a war which the Soviets 'won', and they could not be replaced by other supplies. In recent years the United States has come to dominate the world grain trade in a manner that far outweighs the Saudi Arabian domination of oil. It now supplies 55 per cent of world exports of grain.

But this powerful position has been bought at a price. The intensification of agriculture necessary to meet this rapidly rising demand is leading to a rate of soil loss that is undermining productivity on one third of US croplands. About 4 per cent of all US land now in crops is losing soil so fast that unless it is taken out of crop production altogether it will soon become worthless. As Lester Brown, a leading American environmentalist, pointed out, 'US farm exports to the Soviet Union amount to a subsidy of Soviet farm inefficiency paid with US topsoil'.

The advantages for global security of a growing interdependence between the Superpowers are clear. But these political advantages can be gained only so long as the productivity of the ecological systems which underpin agriculture are sustained.

It is easy to forget just how dependent even the most industrialized of countries are on products which are biological in origin. We rely on natural systems for all of our food; for the wood to make furniture, buildings, musical instruments, toys, paper and much more; for leather, cotton, wool and other materials used in the clothing, bedding and furniture industries; for chemicals to be used in the pharmaceutical, medical, perfume and other industries. The list is endless. If the productivity of ecological systems declines then, inexorably, the prices of economically vital products will rise and with them political tensions between and within countries.

For our food we depend on the

health of just three systems: the rangelands on which we graze our animals; the croplands on which we grow plants; and the oceans in which we fish. These sources produce over three and a half billion tonnes of food a year: nearly 30 million tonnes of eggs, 300 million tonnes of potatoes, half a billion tonnes of milk, one and a half billion tonnes of cereals. Yet in the midst of all this abundance nearly one billion human beings live in absolute poverty, 'a condition of life so degraded by disease, illiteracy, malnutrition and squalor as to deny its victims basic human necessities', according to Robert MacNamara. Half of these agonized people eat less than the

Palouse County, USA, in the heart of the North American breadbasket.

'minimum critical diet' each day, or in plain language, are slowly starving to death.

There are only two approaches we can adopt to reduce this ocean of misery. We can produce more food, either by bringing new land into cultivation or by increasing the productivity of land already in use, and we can share the available food more equally. It is no accident that the period of relative political stability since World War Two has coincided with a period of relative food security. Although global population has grown rapidly, agricultural production has grown faster. Thus despite ups and downs and inequitable distribution, the general trend has been for food consumption per capita to rise.

But this era has come to an end. The explosive growth in grain production which saw world production double in the twenty years from 1950 to 1970 was fuelled by the cheap oil that stimulated the massive applications of chemical fertilizers, and by the increases in irrigation and mechanization necessary to achieve these productivity gains. Energy was substituted for land, so successfully that for long periods the Americans could pay farmers to keep land out of production, thus providing the world with a margin of food-producing reserve. We all know that the era of cheap energy came to an abrupt halt in 1973. What is less well known is that the era of food security was brought to an end at the same time.

The prospects for increasing productivity further are not dramatic. In the industrialized countries most of the available croplands are already back in production and intensification has already reached a level at which further efforts produce rapidly diminishing returns. Indeed, as was pointed out above, the ecological degradation resulting from very intensive agriculture may require land to be taken out of production in order to preserve its long-term productivity. In the developing countries, even though current levels of productivity are often low, the

cheap energy necessary for the massive gains of the post-war period is no longer available.

Other, equally intractable, obstacles exist to increasing food production in the developing countries. Evidence from all over the world suggests that small, family-owned farms produce more food than large ranches and estates in most circumstances. Yet in Latin America the FAO found that just 7 per cent of landowners possessed 93 per cent of the arable land. In other countries unreformed tenancy arrangements leave millions of insecure peasant farmers with no incentive to improve their yields or the long-term productivity of their plots, since the benefits have to be shared with unpredictable and often greedy landlords. In many countries, the better valley lands are occupied by affluent individuals or large, frequently multinational, corporations. These estates often produce cash crops or livestock for export leaving the majority of peasant farmers to struggle for a living on steep slopes that should not be farmed at all.

Throughout the world a rising tide of landless peasants is being swept into ecologically fragile, marginal lands. Lacking other means to make a living, land-hungry farmers clear forests that are needed for ecological protection or wood production. In dry lands, farmers are forced to plough areas with unreliable rainfall and erosive soils. This in turn forces herders to squeeze their livestock into smaller areas,

Are these combine harvesters perhaps our best defence against a nuclear war?

leading to overgrazing and the degradation of rangeland into desert. And when, as they inevitably will, the rains fail, the newly ploughed fields degenerate into dustbowls.

Land reform alone will not solve all the problems of increasing agricultural productivity in the developing countries. There will also need to be much greater support in terms of credit, research and development, extension services and the directing of investment into rural areas. But without land reform the vicious spiral of poverty generating ecological destruction which generates yet more poverty will continue unchecked. In the developing countries the politics of ecology are also the politics of social justice.

It is the tragic banality of the hamburger which brings this problem into its sharpest focus (*See also* Chapter Three). During the 1960s Costa Rican beef production almost doubled, yet per capita beef consumption in Costa Rica fell by almost one quarter. The missing beef had gone to feed the insatiable appetite of Americans for pure beef hamburgers. In the process, some two thirds of Costa Rica's natural forests, one of the richest ecosystems on Earth, have been cleared to create artificial pasturelands for the cattle. Furthermore, the cattle ranches are owned by a mere 2,000 ranchers and account for about 51 per cent of the agricultural land in use in the country, with all the consequences for peasant farming described above. This same pattern of exploitation of both human beings and ecological systems is repeated throughout Central America and Amazonia.

But the Americans are not alone. In every one of the developed countries, rising affluence has brought with it a rising demand for beef and other meat products. To meet this demand cattle are fed grain that is fit for direct human consumption. This grain is converted very inefficiently into animal protein; typically between 2 and 4 kg of grain are needed to produce 0.5 kg of meat. Thus the per capita consumption of grain in the developed world is fast approaching one tonne per year with only about 10 per cent consumed directly as grain. In developing countries, almost all the 160 kg per year consumed (in places like India, China or Kenya) is eaten as grain.

The world harvest of grain – currently about one and a half billion tonnes a year – is enough, equally distributed, to give everyone about 1 kg (two pounds) a day, well above the minimum needed to sustain health. There is, however, no global mechanism for ensuring equitable distribution. Thus the rich countries, with incomes typically twenty times greater than the poor countries, are able to compete more vigorously in world markets, as the Soviet Union did in 1976, for grain to enrich diet rather than to ensure survival. As long as agricultural productivity was rising very much faster than population growth this was unfair but not catastrophic for the poor. If the era of great increases in production is now over, at least for the foreseeable future, then the wealthy can only enrich their diet by taking food from the mouths of the poor.

What is most bitterly ironic about this prospect is the growing evidence that this same shift in diet is the fastest-growing cause of death among the rich. The diseases of affluence, heart attacks and cancer in particular, are in part a product of an over-rich diet. The failure of our political processes can have no more telling mark than that given a planet capable of producing enough for all, we approach the end of the century with 20 per cent of the world's population in danger of dying from undernourishment and another 20 per cent in danger of dying from over-eating.

The Politics of Forests

On a morning in March 1973, in the remote hill town of Gopeshwar in the Chandi district of India, a group of workers arrived from a sports goods factory. They had a contract from the forest department to cut ten ash trees in

199

the nearby village of Mandal. The villagers were opposed to the cutting of the trees and said so. When the workers insisted on felling them anyway the villagers hit on the idea of hugging the trees to protect them. With this act was born Chipko Andolan – the movement to hug trees – a movement that has helped draw global attention to the plight of our forest lands. In India, as in many other countries, the politics of ecology is the politics of protest. In this case it is protest at policies which have seen 4 per cent of India's remaining forest lands lost to agriculture in less than twenty years, reducing forest cover in India to just 12 per cent of the land area.

Forests are the fourth of the great ecosystems on which we depend for so many products and services. Forests play an essential part in the global cycling of carbon, oxygen, water and nitrogen on which all life depends. They also help to stabilize water systems and thus to prevent flooding by slowing and evening out the rate at which rain runs into rivers and streams. In the industrialized world, wood is essential for construction and furniture, to provide paper without which education, commerce and government could not function, and to supply gums, resins and oils to industry. Even in the United States firewood still supplies more energy than nuclear power stations. In the developing countries the dependence on forest products is much greater still. Wood is often the only source of fuel and the only building material; most farm implements are made of wood; medicines and dyes and a significant amount of food from wild fruit, nuts, honey and game are also taken from woodlands.

Yet despite its central economic importance, per capita production of wood has been falling since 1967, as shrinking forests and growing populations have generated a demand gap bridged by rapidly rising real prices. Tropical moist forests, among the most biologically productive – and fragile – of forest lands, are being cut at a rate of fourteen hectares

a minute, destroying an area the size of Hungary each year. In addition, millions of hectares of open woodland are being cleared or progressively degraded by overgrazing, clearance for ploughing, and firewood collection. Many governments find it easier to encourage, or turn a blind eye to, new settlements on forest land as an alternative to the political difficulties of land reforms that would reduce land hunger. Rising oil prices have forced the replacement of kerosene (paraffin) with firewood in many developing countries; one quarter of the world's population now lives in areas where the collection of wood outpaces new growth, thus depleting the resource. Unregulated timber felling, largely by giant multinational corporations, effectively treats forests as 'mines' to be worked out as rapidly as possible, rather than as renewable resources to be farmed on a sustainable basis. And in the wake of the foresters, whose roads open up new areas, come the settlers to clear the remaining trees.

There is no shortage of answers to the problem of deforestation. Foresters, soil scientists and farmers often know exactly what needs to be done in order to reverse these dreary trends. What is missing most often is the political will to implement solutions. Where the political will is present the results can be remarkable. This is true whether the economy in question is centrally planned, as in China, or a free market, as in South Korea. In both countries, contrary to the trend elsewhere, the area under forest has increased dramatically in recent years, in China from 5 to 12.7 per cent and in South Korea by nearly a million hectares. These successes, and those of similar ventures in other countries, have all shown that neither the individual villages, nor the market place, nor the government, nor the experts, working alone, can hope to overcome the complex problems of ecological degradation. It is only when they all work together that success is achieved. Thus, the politics of ecological recovery is the politics of cooperation.

The Politics of People

'Despair', said C.P. Snow, 'is the only sin.' There is a great temptation, faced with the unfolding complexity of the ecological crisis in which we are already engulfed, to despair; to feel that our own ideas and actions are inconsequential, that the combinations of circumstances that might lead to a sustainable future are so unlikely as to be unachievable. Commentators often seem perversely fascinated by the sheer scale and difficulty of the problems facing the planet; to discuss them as one might discuss the heat of the flames or the depth of the water on a ship that was burning and sinking.

It is a common mistake to look at the length of the road ahead and to forget the distance already travelled. It is equally common to feel that one is facing these problems alone. Both mistakes are recipes for despair.

People are the only cause of ecological devastation on a global scale, but they are also the only source of its recovery. What has been most remarkable since the environmental crisis exploded into human consciousness has been the speed and depth of the response. It is very easy to dismiss the decade-long series of United Nations Conferences on specific environmental problems as mere talking-shops, or to regard the long list of international conventions on environmental management that have been agreed in the last ten years as pious expressions of good intent. There is truth, and disappointment, in both perceptions. Nevertheless, it remains equally true that governments have engaged in a ten-year debate on the future of the global environment, a historically unprecedented occurrence. And if nothing else has been achieved, the enormous increase in understanding of both our environment and each other that has occurred has made the effort worth while.

Just how powerful the unifying force of the environment is, internationally, can be best illustrated by one incident. During the negotiations on the Mediterranean Action Plan (a UN-sponsored plan to protect the ecology of the Mediterranean), official delegates from Egypt and Israel continued to sit round the same table reaching agreements on joint actions despite the fact that their two countries were at war at that moment.

Of course it remains right to feel impatience and frustration at the slow pace of international response compared to the accelerating rate of destruction, but it is equally important to realize just how rapid, in political terms, the international mobilization on the environment has been. Impatience and frustration are also thoroughly justified at the inertia of governments at national level. Much has been achieved, huge bodies of national law that simply did not exist ten years ago are now to be found on statute books throughout the world. Institutions to manage resources and control pollution now exist where there were previously none. But all too often, despite the consistently high level of political support for vigorous environmental action in the polls, laws remain unimplemented, agencies unfunded.

It is a measure of the impatience felt by many people that there are now ecological parties in at least twelve countries in Western Europe alone. In Germany and Belgium they have elected representatives in their national parliaments and in many other countries they are represented on local authorities. What all of these parties have in common is a clear commitment to place the environment at the heart of the political process, rather than consider it a remote appendage as do the orthodox political parties. Although the political and environmental conditions which stimulated the formation of Green parties vary widely from country to country, a common feature throughout the world has been the total failure of both the dominant political traditions to incorporate ecological imperatives into their thinking.

Neither the free-enterprise economies of the West, nor the centrally planned economies of the East, have made much progress in establishing equilibrium between ecological and economic needs. Despite the great variety of ways in which these dominant traditions manifest themselves in the world there is not yet one country in which short-term, often self-defeating, economic criteria do not always outweigh longer-term ecological criteria in the choice of policy.

This should be no great surprise. Politics is nothing more than the process by which we make choices. To the extent that we agree with, condone, or simply ignore the choices that are made on our behalf, we get the politics, and the world, we want. Most men and women lead lives of small satisfactions in which their primary preoccupations are with their family, their home and their job. These limited horizons mattered less when their choices were of little global consequence. But they matter a great deal now that we live in an interdependent world with vastly enhanced powers. When I choose to eat a hamburger in Britain I am quite literally taking food from the mouth of a hungry child, when I turn on an electric switch I am choosing to dump acidic rain on the forests and lakes of Sweden, as I throw away a cardboard box I am choosing to cut down more of our shrinking forests.

The task facing a legitimate politics of ecology is twofold. Firstly, it must engender in each one of us a sense of personal responsibility for the fate of the planet – an awareness that our personal choices, no matter how apparently trivial, when aggregated together are what determine that fate.

The second task is that of developing and articulating a political vision that will shape the individual choices of millions upon millions of individuals so that the sum of all their individual choices is the voluntary choice of an ecologically

·sustainable world. Ultimately, the environment cannot be defended by laws, no matter how strict the laws or powerful and rigorous the governments that enforce them. Laws, and the means to enforce them, will always have their place in protecting the environment. But no code of law is proof against the contrary choices of billions of individuals. It is only by the positive affirmation of the mutual interdependence of the ecology and the economy, of Man and Nature, by individuals through the political process that we can hope to achieve the equilibrium necessary to avoid the looming catastrophes. It is the task of the politics of ecology to make that affirmation.

Tom Burke

Dieter Drabiniok and Gert Jannsen, two of the 27 members of the Green Party elected to the German Parliament in 1983.

*There are over 10,000 organizations actively
campaigning to protect the environment, many of
them formed in the ten years since the Stockholm
Conference..*

Chapter 12

Environmental Hot Spots

By the 1980s the international character of many environmental problems had become clear. It has been the theme of much of this book that the effects of deforestation, desertification, acid rain, the loss of wild species and the serious degradation of the coastal marine environment cannot be contained within national boundaries, nor can they be effectively dealt with exclusively at the national level. In some cases, countries under serious threat simply do not have the resources and expertise to deal with the problems. In others, the problems spill over national frontiers, and co-operative, international effort is required to limit the global impact of serious natural resource depletion and unsustainable economic growth.

Ecological danger signals in places as different and far apart as Scandinavia and Africa, Nepal and the Caribbean Islands have caused environmentalists to turn their attention to the wise management of the Earth's renewable resources, and to the global effects of industrial pollution and the disposal of the deadly by-products of industrial processes. If the survival of Man's life-support systems – the natural world of plants, animals, soils, waters and air, upon which the whole complex structure of human society and welfare is built – is to be assured, conservation and economic development have to go together.

At the beginning of the decade, warnings of the long-term implications of present environmental trends, contained in reports like the United States' *Global 2000 Report* and the *World Conservation Strategy*, made it clear that in Man's battle against nature, not only was nature losing, so was Man. (*The Global 2000 Report to the President*, Council on Environmental Quality, 1980, Washington DC; *World Conservation Strategy*, International Union for Conservation of Nature and Natural Resources (IUCN), 1980, Gland, Switzerland.) Unrelenting pressures on a finite resource base, our fragile 'small planet', would eventually impoverish the quality of life for people everywhere by the end of the twentieth century, even in spite of greater material output, unless all nations acted together to change these trends.

Attention has shifted dramatically away from the pollution and energy problems of the industrialized countries to international environmental issues and to the population and natural resource management problems in the developing countries of the Third World: Asia, Africa and Latin America. Here, on these three continents, live three quarters of the world's four and a half billion people. And here is where 90 per cent of the population-increase, projected to be a further two billion by the year 2000, will occur (*See* Chapter Two). Most of these people are poor, struggling to survive. Most of them live directly off the natural environment – unable, because of their poverty and growing numbers, to replenish the Earth or temper their demands on limited, fragile ecosystems. In the words of environmentalist Erik Eckholm (in *Down to Earth*, 1982) the growing populations of these countries are 'caught in endless cycles of hunger, illiteracy, exploitation, and disease (and) have no time to worry about global environmental trends. Many are forced by circumstances beyond their control to destroy the very resources from which they must scrape their living.'

Interdependence has become the environmental as well as the economic

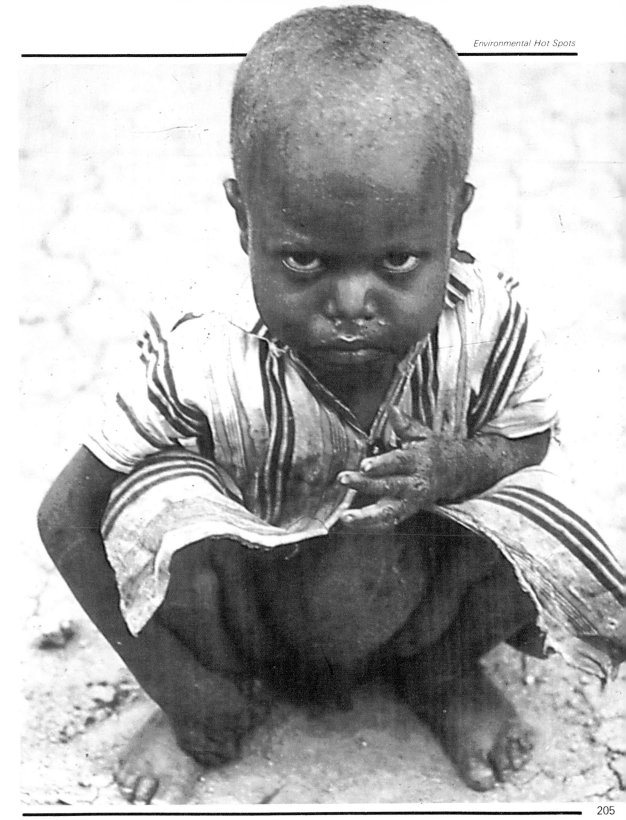

In the time it takes you to read the page opposite, three young children will have died of malnutrition and preventable diseases.

GLOBAL HOT SPOTS

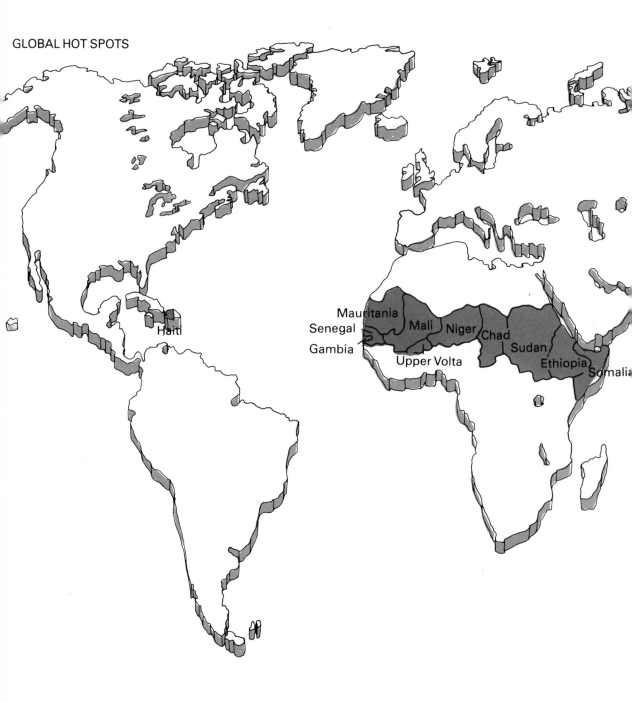

Mauritania
Senegal
Gambia
Mali
Niger
Chad
Upper Volta
Sudan
Ethiopia
Somalia
Haiti

*In a growing number of countries, environmental
degradation is proceeding at such a pace that it is
already beginning to undermine their capacity to
support their population. These countries offer us an
early warning of what will happen if action is not taken
to halt environmental damage.*

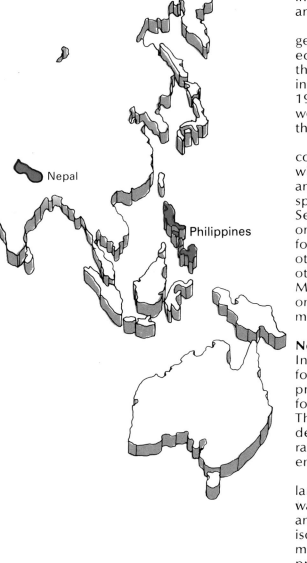

Nepal

Philippines

reality of the 1980s. If environmental disaster in the year 2000 is to be averted, global action is needed now. Armed with vastly increased scientific knowledge about dangerous ecological trends, environmentalists have been turning their attention to the centres of political power in both the rich countries of the North and the poor countries of the South.

It is here, in countries still poles apart geographically and culturally, but economically and environmentally linked, that support for concerted national and international action has to come. For by 1980, poverty and expanding populations were already pushing some countries to the brink of ecological disaster.

It is not possible to catalogue every country under severe ecological threat within the bounds of a single chapter, but an account of some of the worst 'hot spots' – devastated areas – can be given. Selectivity has demanded concentrating on areas each of which, in an extreme form, illustrates global problems affecting other countries or regions. Clearly many other countries are threatened – Madagascar and the Ivory Coast to name only two – but the following examples must stand for the whole.

Nepal: A Mountain Kingdom Under Threat
In the past, Nepal was covered with forests. If trees continue to be cut at the present rate, however, all of Nepal's forests will be gone by the year 2000. Threatened by deforestation, soil erosion, declining agricultural productivity and rapid population growth, Nepal's environment is at breaking point.

The kingdom of Nepal is a small, landlocked Asian country. Until 1951, it was ruled by the hereditary Pana dynasty and was virtually closed to outsiders. This isolation has given Nepal a sense of mystery and charm that belies its critical problems of development. Nepal is dominated by its mountains, the Himalayas. They are the world's youngest, highest and steepest mountains, composed of fragile sedimentary rock that

enforces on Nepal one of the world's more delicate ecosystems. The country is subject to powerful natural erosion processes because of the steepness of the terrain and because most of the annual rainfall comes in torrents during the summer months.

Wedged, long and narrow from east to west, between India to the south and Tibet and China to the north, the Himalayan range separates tropical south Asia from the dry Asiatic plateau to the north. Nepal's altitude along its long southern border with India is nearly sea level. The mountain peaks to the north average over 20,000 feet in height. This steep north-to-south drop contributes to Nepal's major environmental problem: soil erosion and declining agricultural productivity.

Wind and rain erosion of the steep mountain slopes is a natural process that is continually breaking the mountain rock down into soil, which is then carried away in Nepal's many swiftly-flowing rivers to the alluvial plains of the Terai.

But increasing numbers of people now account for more erosion than that which occurs from natural causes. Sir Edmund Hillary, who knows the Himalayan region as well as any outsider, has given a vivid account of the results in Chapter One. Expanding populations require ever more land for agriculture and ever more wood for fuel.

Ninety-five per cent of Nepal's fourteen million people are farmers who live in the hills, in the middle belt of the country. The ratio of people to cultivated land is extremely high, with the densest

Machha Puchhave, Annapurna Range, Nepal. Nepal consists essentially of three zones: the mountains, averaging over 20,000 feet; the hills, and the plains, which account for only ten per cent of the land area. Most of Nepal's population is confined to the hills.

populations living in the hills. On a land-base roughly the size of Florida, Nepal's total population is expected to reach 22 million by the year 2000, an increase of 58 per cent in twenty years.

Nepal's forests are suffering as a result of these pressures. They are the country's most valuable resource, and the most abused. They provide fuel, fodder and building materials. Just as importantly, they guard against destructive soil erosion. The protective and life-maintaining

and down, instead of along the contour, and they plant the same crop year after year, leaving little other protective vegetation on the fields and not allowing the soil to lie fallow and regenerate itself.

Deforestation and a high population density per unit of arable land, combined with natural wind and rain erosion and erosion from the overgrazing of animals, make Nepal's prospects for the future very bleak.

But they are not hopeless. If action is

functions of Nepal's forests cannot be overestimated.

Even though Nepal's economy is based on agriculture, only 10 per cent of the total land area is flat or gently sloping, genuinely suitable for farming. As a result, forests in the hills and on steep mountainsides have been cleared for farming. The need for agricultural land and fuelwood, and the grazing of animals have caused not only serious soil erosion in the hills, but also floods and silting-up downstream in southern Nepal and in India. Two hundred and forty million cubic meters of topsoil are carried away annually, severely limiting Nepal's agricultural productivity.

Traditional farming techniques practised by the Nepalese exacerbate the soil erosion. Nepalese farmers plough up

taken soon, the trend can be reversed. Nepal's mineral resources have yet to be exploited. The country's many rivers have tremendous potential for the development of hydroelectric power and irrigation schemes. The technology and information now exist to enable much of the soil erosion from both farm and forest lands to be controlled. Proper farm practices can reduce soil loss by between 50 and 90 per cent. Planting trees, the maintenance of grass and cover crops on the exposed ground, and controlled grazing can greatly reduce losses on steep slopes. But the people of Nepal, who have an appallingly low life-expectancy of 44 years, are far removed from Western society and its techniques, and suffer from poor health, few educational facilities and almost no economic development. They

Traditional farming in Nepal.

need to be shown these modern techniques if they are to be able to help themselves. Living in isolated hill villages, farming as their great-grandparents used to, their struggle to produce enough food for their families on deteriorating land limits their long-term view to the end of each day. Public education on what such a high birthrate means for the family and society as a whole, and easy access to contraceptives are also crucial.

The country's arable land is already densely populated and has sustained much environmental destruction, some of it probably irreversible. The vicious circle of poverty and population growth must be broken if the pressures on Nepal's land and forests are to be relieved, and if the wellbeing of its people is to be assured.

Haiti: an Ecological Wasteland

In the Caribbean, far from the Asian mountain kingdom of Nepal, another nation is in dire environmental trouble. Once called the pearl of the Antilles, Haiti is losing the capacity to feed and provide a living for its five million people. It is suffering from a degree of environmental degradation almost without equal in the world. The cause of this environmental disaster is extreme deforestation, soil erosion, poor farming practices on unsuitable land, and overgrazing.

In 1979, a United Nations official predicted that Haiti 'may well become the Caribbean's first true desert'. A 1980 US Government report predicted gloomily that, unless present trends were reversed, 'the country will be an ecological

Deforestation in Nepal: the remains of the forest.

wasteland by the year 2000'.

Haiti's natural resources are being destroyed to the point where rehabilitation of the resource base may soon no longer be possible. Because the economy is based on agriculture, that too is being ravaged as the productivity of its soils declines.

Once a land rich in natural resources, now its people suffer from hunger and malnutrition as disease, drought, soil erosion and scarcity of building materials and fuelwood take their toll. The collective effects of Haiti's environmental problems are placing enormous strain on the country's rural population, which makes up three quarters of Haiti's total population. Their average annual income has fallen to 50 dollars. Life expectancy is

51 years. Ninety-six per cent of the people cannot read.

The republic of Haiti is located on the western third of a large Caribbean island that it shares with the Dominican Republic. Almost completely covered in forests when Columbus visited it in 1492, and later the source of food, spices, sugar, coffee and tobacco for its colonial owner, France, Haiti today is the poorest country in the Western Hemisphere.

Today there are few forests left and, if there is no replanting, they will all be gone in less than ten years from now.

What went wrong on this island paradise? Haiti is a country of mountains and hills. Of its total 850,000 hectares (2.1 million acres) of land considered suitable for farming, slightly more than half is in mountainous areas. Expanding population and its growing need for farmland and fuelwood cleared the land of its trees. Deforestation then caused soil erosion and declining agricultural productivity. On the country's steep, deforested hills, there is nothing to stop the soil from being washed away. Heavy rainfalls cause flooding that fills even the streets of the capital, Port-au-Prince, with mud. Eroded soil clogs irrigation works and silts up dams. The country's main source of power, the Peligre dam, has been silted up so quickly that its effectiveness has been reduced by an estimated 40 to 60 per cent. According to a US Government report, 'Deforestation and the attendant erosion of the soil resource is the most basic environmental problem in Haiti. Unchecked, it spells permanent doom for small-farmer agriculture and much of Haitian rural society.'

The social and economic effects of environmental degradation are causing deep and pervasive problems in Haitian society. Tens of thousands of Haitians are leaving their homeland for other Caribbean islands and for the United States, in search of food and jobs.

As in other developing countries, behind Haiti's severe environmental problems lies the population problem. Haiti has the highest population density in Latin America. With 80 per cent of its nearly five million people living in rural areas, this population is further concentrated on the country's arable land. Here population density averages 490 people per square kilometer, which means less than one fifth of a hectare (half an acre) for a family farm. Because there are so many people living off the land, marginal, easily erodible land is farmed and forests are cleared on steep slopes to provide agricultural land and firewood, which is the main source of fuel for cooking and heating in Haiti.

Traditional farming techniques in Haiti are not suitable for sustained production on these marginal lands. They aggravate soil erosion and loss of

Erosion in the Kenscoff area of Haiti.

nutrients. This process of degradation results in a chain of events that leads to more and greater degradation. The poor farmers must use every available resource at hand just to survive. Even vegetable wastes are burned for cooking fuel as trees become increasingly scarce, and the soil loses a valuable source of nourishment. Farmers with small hillside holdings use all their land for crops, so none of the land is allowed to lie fallow and regenerate its fertility by natural processes.

The environmental situation is at such a critical point that action, if it is to have any chance of success, must begin now. An overwhelming problem to be faced by anyone attempting to help is the extremely limited amount of trained manpower in Haiti. Hand-in-hand with conservation must go basic educational and training projects if the process of degradation is to be reversed. Individual farmers need to be made aware of the critical nature of their country's environmental problems. In the past, most conservation and soil preservation schemes have failed dismally because the farmers were being asked to set land aside from food production for tree planting that would stabilize the soil and make things better – in the future. But poor farmers cannot feed their families on long-term benefits.

With Haiti's population increasing at an annual rate of 2.4 per cent (most Western countries are at zero population growth), time is running out.

The Sahel: A Fragile Ecosystem Under Threat

A combination of unwise environmental practices is threatening the livelihoods of millions of some of the world's poorest peoples who live on the arid and semi-arid lands of the Sahel and the Sudanian region. The six main Sahel countries are Chad, Mali, Mauritania, Niger, Senegal and Upper Volta. The countries most under threat in the Sudanian region are the Sudan, Ethiopia and Somalia.

The attention of the world was first turned to this huge region south of the Sahara desert in Africa in 1972 and 1973, when four years of drought ended in catastrophe. Hundreds of thousands of people and millions of their cattle died of starvation during that 'great drought', which has been described in some detail in Chapter Four.

Now, a decade later, the same ecologically unsound development, exacerbated by growing populations of people and cattle, is pushing the same region to the brink of another disaster.

Desertification has been occurring throughout human history, in countries the world over. The Sahara itself was wet and fertile for 3,000 years. But it is only in the Sahel that the economic viability of an entire region, and millions of people, are directly threatened by this process which changes productive crop and rangeland into barren desert. Because many Sahelian countries are among the poorest in the world, the need for international help is especially acute. The World Bank, in a special report on the problems facing sub-Saharan Africa, said a doubling of economic aid in real terms by the end of the 1980s was essential for these countries.

One third of the Earth's land surface is arid or semi-arid. In spite of very low rainfall, arid lands can be surprisingly productive, and skilled herdsmen and farmers can coax a living from fragile

Over the years, the annual per capita income of Haitians has fallen steadily to its present level — $50.

desert environments without over-using them, if their numbers are kept in equilibrium with what the land can sustain. Arid lands are home to about 700 million people worldwide. Eighty-one million live in the Sahel and Sudanian regions.

The United Nations Environment Programme, headquartered in Nairobi, Kenya, has estimated that desertification on the fringes of the Sahara is spreading at a rate of 1.5 million hectares (3.7 million acres) a year. The six main Sahel countries of West Africa, and Sudan, Ethiopia and Somalia, to the east of the Sahel, are losing precious land at the very time when more land is needed to feed their rapidly growing populations. The current annual

increase in population in the Sahel region is 1.5 million people a year. The nomads, who have traditionally grazed their herds on the fragile arid lands, are increasing their population as fast as the farmers and city dwellers in that region.

For centuries, the Sahel and the Sudanian zone have been the domain of nomadic herders. Driving their sheep, goats, cattle and camels across the region in search of grazing was a logical response to the land, the seasons and the pattern of rainfall. When the rainy season produced a brief flush of grasses in the arid zone, the nomads were there to benefit from it. This strategy also removed the herds from the river basins during the flooding season when disease, carried by the tsetse fly, was most prevalent. When the crops were harvested in the river basins, the nomads moved back to graze on the crop stubble and give, in return, fertilizer to the farmers' fields in the form of their cattle's manure.

This pattern was interrupted by the era of colonialization, when national Governments tried to restrict the nomads' movements and keep them to predictable paths. To encourage them to settle down, the Governments began drilling deep wells to provide a permanent source of water for the nomads' herds. This policy had the unexpected result of causing the numbers of animals grazing in the Sahel greatly to increase between 1945 and 1967.

Now, as reports of drought in this region begin to come in again, the imbalance between Man and his environment once more becomes precarious. Across the Sahel danger signals are heard.

In Ethiopia, a new failure of the rains during the past four years has created a new class of environmental refugees. Ethiopia's rapidly growing population is putting tremendous pressure on the land. Its 30 million people are increasing by 2.5 per cent, or 750,000 people, a year. If this rate of increase continues, the country's population will reach 50 million by the year 2000. This growing population and deforestation over wide areas of the country have led to the cultivation of marginal, high-altitude land previously covered by forests. These highlands are losing one billion tonnes of topsoil a year, according to UN estimates. In the worst-affected areas, the Ethiopian Government is moving tens of thousands of its environmental refugees from their over-used, eroded highlands to the relatively underpopulated – because malaria-infected – lowland plains. The highlands, especially the provinces of Wollo, Tigrai and Gonder, have been so overfarmed, overgrazed and deforested that continuing efforts to scrape a living from the area threaten to destroy the land permanently. Now it is feared that the failure of the 1982 June to September rains could precipitate a crisis in Ethiopia as bad as the great drought of 1972–1973.

In the Sudan, cutting trees for fuel and clearing land for agriculture has destroyed large areas of the country's natural forest. Over 500,000 square kilometers of forest have disappeared. Sudan's deserts – which cover an area slightly larger: 650,000 square kilometers – are extending on the other hand at a rate of five to six square kilometers a year. Since the mid-1950s, Sudan's desert has crept 100 kilometers southward from the Sahara, reducing the land under cultivation in the Northern Region.

As the population grows, without a corresponding transformation of agricultural and livestock-rearing techniques, the traditional cropping and grazing cycles, which are sound and sustainable when followed properly, break down. In Sudan's northern province of Kordofan, livestock numbers have multiplied six-fold between 1957 and the mid-1970s, limiting the amount of land available for crop production. This has happened during a period of time in which declining soil fertility and falling crop yields meant that more land was needed just to produce the same amount of crops. In 1973 in this province, five

Top: most of Haiti's once verdant forests have already gone.

Above: not surprisingly those who can leave, do so. Haitian boat people.

hectares were needed to produce the same yield of peanuts (groundnuts) that had come from *one* hectare in 1961.

In Senegal, the most westerly of the West African Sahel states, environmental problems are at such a critical level that, if they are neglected, according to a report by the US Agency for International Development, they will seriously impair, if not make impossible, any economic progress in the country.

A quarter of Senegal's territory is arid land, and 70 per cent is semi-arid, which means that nearly all of the country and its five and a half million inhabitants are threatened by desertification. The most critical areas are in the northern part of the Fleuve region, Senegal-Oriental, parts of Casamance and the Cap-Vert region. As in other arid regions, Senegal's problems are being caused by uncontrolled grazing and growing numbers of cattle, the overcutting of trees, and the over-cultivation of arable land. But bushfires are also a problem in Senegal. Either accidental or set to clear land for agriculture, they burn over 40 per cent of the country each year, wasting forest and pasture. The forests are also under heavy pressure to provide fuelwood and charcoal for domestic cooking in the rural and urban areas. Most Senegalese, even in the cities, rely completely on wood for their fuel needs.

Senegal's growing population has forced agriculture onto fragile, marginal lands. It has also obliged lands already under use to become more intensely cultivated and so become susceptible to erosion by poorly adapted new techniques of cultivation. Irrigation techniques have been introduced without giving farmers adequate training, and the result is salinization and waterlogging of the soils, causing the land to be abandoned. Under traditional cultivation, soil fertility is also diminishing because fallow time is reduced or dispensed with altogether. Then, when these fields are abandoned in times of drought, the unprotected soils are open to devastating

erosion.

Mali, bordering Senegal to the west and Mauritania and Algeria to the north, is also dominated by an arid and semi-arid climate. Nearly half the country is covered by dry savannah. The southward expansion of desertification from the Sahara has hit Mali very hard. Social patterns and the livelihood of the nomadic peoples in the savannahs have been disrupted. In addition to suffering from soil erosion, deforestation and rangeland degradation, malaria and the debilitating disease known as river blindness are major health problems for Malians.

Because of the serious threat posed by the process of desertification to the wellbeing of the peoples of the Sahel, international and bilateral aid agencies

Nomads in the Sudan. The deserts in the Sudan are creeping southwards at an inexorable 3 kilometers a year.

have given high priority to the environmental problems of this region, since the 1972–1973 drought. In spite of this, progress in halting desertification has so far been slow. Many schemes have been discussed and tried, but visible signs of progress are few (See Chapter Four).

Although experts disagree whether the next crisis is just around the corner or some years away, no one denies that if present trends continue it must come. Population in the region is growing at an annual rate of 2.5 per cent, but food production only increases by one per cent a year. Foreign exchange, and there is very little of it, must be spent on importing food, which further reduces the region's ability to invest in rural, agricultural development. Without this kind of investment there will be no economic

progress and the fragile environments may be pushed to the point of no return.

Can the spread of desertification in the Sahel be halted, or should these countries be seen as environmental write-offs? Because desertification is a man-made phenomenon, its control is theoretically in the hands of men. The greatest losses to desertification are not due to inexorable spread of true deserts, but to misuse of rangeland and farmlands. While it is true some arid lands have been destroyed beyond the point of repair, most desertified lands could be restored if the correct measures were undertaken. And many of these semi-arid lands, like the people of Sahel, are incredibly resilient.

With the grass cover gone there is nothing to stop the wind stealing the soil.

The Philippines: The Destruction of Coastal Marine Ecosystems

The Philippines is a long archipelago, made up of over 7,100 islands. The coastline is 16,000 kilometers long – twice as long as that of the continental United States. Its coastal marine environment is extremely important for its livelihood. The estuaries, mangrove swamps and forests, salt marshes and coral reefs of this tropical country are rich in aquatic life. Over 750 species of saltwater fish are used for food. Shellfish, including crabs, shrimps, prawns, clams and oysters, are an important export for the country as well as a source of protein for the people. But these complex ecosystems are being rapidly destroyed.

Half of the coral reefs fringing the Philippine islands, on the south-east rim of Asia, are already dead or dying. Half of the country's mangrove forests – which used to cover 400,000 hectares (over 1,500 square miles) – have been destroyed. The remaining mangrove forests are being cut down at a rate of 24,000 hectares a year. If that rate continues, there will not be any left by the end of the 1980s.

Coastal areas are crucially important sites for two thirds of the world's fisheries. They are highly productive biologically and support a wide range of economic activities. In the Philippines – and in many other countries, including the United States, Sri Lanka, Thailand, Malaysia and the Caribbean islands – such coastal areas are severely threatened by other kinds of economic development. Factories, refineries, housing developments, power plants and port facilities are taking a rapid toll. Industrial pollution and raw sewage are emptied into the coastal waters. Increasing numbers of people and the need for export earnings lead to over-exploitation of the fish and shellfish.

For the country as a whole, fish provide an important source of protein. Because malnutrition is serious and widespread in the Philippines, especially for children, the loss of the productive marine ecosystems would pose grave problems.

Mangroves have always fascinated scientists. Hugging the shore, they grow in salt water and have unusual ways of adapting to their environment. They produce seeds that germinate on the tree. They have specialized 'prop roots' that grow out like fingers to keep themselves steady against the waves. Their leaves have salt-excreting glands. Like coral reefs, they physically protect both marine species and the land itself. Unlike coral reefs, they are not tourist attractions, and are often thoughtlessly destroyed by developers. It is only recently that mangroves have been discovered to be

Philippines: mangrove swamp.

very productive ecosystems.

Their leaves, when they fall, serve as the base of a food chain that supports most of the coastal commercial fisheries. The leaves fall and decompose, and are fed on by fungi and bacteria. These are then eaten by fish and shellfish. Mangrove roots also provide nurseries for fish and shellfish, and they help regulate water quality by processing sewage, absorbing the nutrients. They also protect the shore from storms and floods.

Coral reef ecosystems are another highly productive marine environment. Coral reef habitats are as complex and as

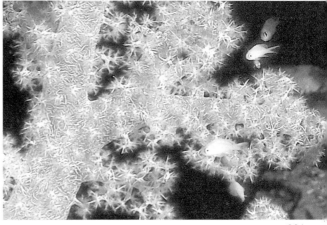

Philippines: soft coral.

Left: crumbling infrastructure. Throughout the industrial world, the civil engineering structures, on which life in our cities depend, are beginning to collapse. Water and sewer pipes fracture daily. Bridges are closed, buildings shut, roads diverted, as we fail to keep pace with a mounting tide of repairs. In the United States a quarter of a million bridges have serious structural problems. In Britain, seven million houses are in need of repairs costing over £30 billion. The collapsing sewer system, into which the lorry shown in this photograph fell, is threatening to undo the achievements of 30 years' work to clean up our rivers. Tens of billions of pounds will need to be spent in the next two decades to ensure that our civic infrastructure does not collapse.

Above: noise is perhaps the most endemic and most intractable of environmental problems facing the industrialized world. Noise imposes stress, disturbs sleep, reduces efficiency at work, and causes severe annoyance to people. No one has yet satisfactorily quantified the economic costs of noise pollution, but they are likely to be very large indeed. The growth of traffic on the roads and in the air has been the most significant source of exposure to unacceptable noise levels. Sudden and unexpected noises can produce profound physiological responses, especially among the very young, the sick and the elderly. Complaints about noise are one of the fast-growing problems facing pollution prevention officers. Furthermore, occupational noise is an important cause of deafness. Over 600,000 people in Britain alone are currently suffering progressive hearing degeneration as a result of exposure to noise at work.

Left: some 60,000 synthetic chemicals are produced commercially and 1,000 new ones added to the lists every year. Our daily lives are now thoroughly permeated with the use of chemicals. The benefits of these compounds are clear. But an unknown proportion cause cancer, birth defects or other human ills. They also damage animals and plants. Because we accepted the growth of chemical use without much control in the past, we now face the expensive and complex task of identifying those chemicals in use which are dangerous. In 1980, a US government report described the problem as: 'staggering in view of the number of substances whose risk should be evaluated'.

Above: nuclear Power Plant, San Onofre, California. The accident at the nuclear power station at Three Mile Island in March 1979 focussed global attention on the dangers of nuclear energy. Just under 200 nuclear power stations are now in operation in some 20 countries. Plans to extend this capacity to over 500 stations in 36 countries have been cut back as costs have risen steeply and public opposition deepened. Although, when operated properly, nuclear power stations themselves release very little radiation, a serious accident could release huge quantities of radioactive materials and render the surrounding countryside radioactive for many decades. The radioactive waste produced by all nuclear power stations, however, poses a much more difficult environmental problem and one for which no acceptable solution has yet been found. Furthermore, the mining of the uranium fuel for nuclear power also creates difficult, and sometimes intractable, environmental hazards.

diverse as tropical rain forests (*See* Chapter Five). They provide food and shelter for many fish and shellfish species, and they protect the land from the ocean forces. In the Philippines, where there are thousands of kilometers of coastline, the reefs help keep the islands from being washed away. The living organisms that make up the reefs also have medicinal properties that are only now beginning to be discovered. These include antibiotic compounds in algae, anti-virus compounds in sea squirts, seaweed and sponges, anticoagulants in the sea snail, and anti-cancer compounds in corals. All these are of potential economic benefit to the Philippines.

The Philippines are fringed by nearly 27,000 square kilometers of coral reefs. Many of them are supposed to be protected as national parks, but they are in danger of destruction from many quarters. Fishermen use dynamite near the reefs to stun fish and to break up coral, which is used for jewelry and in the construction trade. Silt from dredging and from eroded, deforested slopes is carried by rivers and deposited in the lagoons and on the reefs. Silt, which smothers the delicate, slow-growing coral polyps, is the most serious threat to the coral reefs' survival.

The pressures of population growth and industrial development have created serious environmental problems in the Philippines. In a country whose actual land area is small – slightly larger than the state of Arizona in the United States – there are about 46 million people. These are expected to increase to 78 million by the year 2000. If the present trends continue throughout the next two decades, the increases in human population density and industrial and commercial activity will increasingly have harmful effects on the Philippines' tropical environment. The first casualties are likely to be the mangrove and coral reef ecosystems.

These environmental hot spots are only the early symptoms of the corrosive degradation to come. They are the result of our past mistakes which are being repeated on a growing scale throughout the world. We have concentrated our attention on developing countries because it is there that the problems are most acute and the means for solving them least available. But the developed countries have hot spots of their own. The four examples illustrated here (see pages 222–5) are just a few of the many serious environmental problems facing the governments and people of the rich world. Each one of these hot spots, and of the many others we do not have space to illustrate, is a warning signal. We will ignore that at our peril.

Athleen Ellington

Chapter 13

The Future

Cultures can collapse. They have done so before, many times. The sands of the Middle East and the jungles of Central America carry the mute reminders of civilizations that ignored ecological constraints and paid a permanent price for so doing. It is hard to imagine the collapse of the world we know, especially when the symbols of its permanence, skyscrapers, motorways, factories, seem so powerful. It would have been equally hard for the Mayans, the dominant civilization of their era in the Americas, to imagine their collapse. Yet collapse it did. In just a few decades a civilization that had lasted fifteen centuries, capable of building cities that have outlived their creators by a thousand years, saw its population crash from five million to half a million.

It would be foolish to suggest the imminent collapse of our civilization. It would be equally foolish to ignore the warning signs. The collapse of the Mayan empire, in common with that of other

227

Chichen Itza, Mexico. Mayan ruins that have outlived their creators by 1000 years.

empires in America, Africa, the Middle East and Asia, occurred when its attempts to sustain a growing population, given the knowledge and technology of the times, undermined the biological productivity of its available land. Until this century, there was no global civilization. Cultures rose and fell in isolation from one another. If one collapsed there was always another to act as a repository for the accumulated experience of mankind. The ascendancy of the European mode of civilization over the whole planet has created an interdependent world in which the fate of each one of us is bound up with the fate of the whole. If this civilization collapses, there is no alternative.

Throughout this book we have been examining the ecological base on which this global civilization is founded. Everywhere, in northern and southern hemispheres; in capitalist and socialist countries; among rich people and the poor; in deserts, forests, oceans; in the very air we breathe, we have found signs of growing ecological stress. Here and there, especially in the more ecologically fragile parts of the planet, there have already been serious local breakdowns in the ability to sustain life (*See* Chapter Twelve). There can be no doubt that the combination of growing population, rising expectations and the enhanced capacity to transform the environment has set us on the same path that has been trodden by other civilizations on their way to collapse. There can be, and is, much argument about how far along the path we are and how fast we are travelling, but the direction is not in question.

Two reports, both published in 1980, provide signposts to our future. The first, prepared by the Government of the United States, examined probable changes in world population, natural resources and the environment through until the end of the century. Its opening sentence was unusually emphatic for an official report: 'If present trends continue, the world in 2000 will be more crowded, more polluted, less stable ecologically and

more vulnerable to disruption than the world we live in now.' (*The Global 2000 Report,* 1980.) The second, prepared by the International Union for the Conservation of Nature and Natural Resources, outlined a strategy for preserving the ecological base of civilization. It identified three objectives that must be achieved if we are to avoid irreversible environmental degradation. These are the maintenance of essential

A washing machine factory: the spread of consumer goods in the West was fuelled by the post-World War II boom.

ecological processes and life support systems, the preservation of genetic diversity and the sustainable use of species and ecosystems. This report, called the *World Conservation Strategy*, then went on to outline the policies that must be followed to achieve these objectives and thus to create a sustainable society.

Choice

What is crucially different for our civilization is that we have a choice. Previous civilizations, faced with ecological collapse, neither understood the nature of their problem, nor were they capable of devising a response. As these reports demonstrate, we can do both. Understanding a problem and its solution is one thing, actually solving it entirely another. Personal and public histories alike can supply plentiful examples of well understood problems, with available solutions, that remained unresolved because the necessary private or political will was lacking. Faced with the scale of change in attitude and behaviour necessary to build a sustainable society it is easy to lose hope. The counsels of despair point to people's apathy, ignorance, greed, selfishness; to a supposed reluctance to change.

Yet the evidence belies these perceptions. In the three decades since World War Two, lifestyles have been transformed. Successive waves of consumer products – television sets, washing machines, automobiles, calculators – have been absorbed into our lives rapidly and painlessly. Habits and customs that had endured for centuries have been swept away in decades as people adjusted to new possibilities in their lives. This capacity for change cannot simply be explained as the pursuit of immediate material satisfaction. Campaigns to halt smoking or encourage exercise have changed the personal habits of millions of people with astonishing rapidity. What is perhaps closer to the truth is that, presented with a clear choice, given good information about its implications, some incentive to choose and a small amount of positive leadership, most people are prepared to change, and change quite rapidly.

Those most aware of the problems are often tempted to repeat their analysis rather than instruct us on what to do next. On a ship that is sinking and aflame, we need to be told where the lifeboats are and which way to row to reach safety, not that the fire is spreading and the sea rougher. All crises have opportunity as the other side of the coin. This is as true of the ecological crisis as of any others. As the pressures have mounted and our knowledge of them deepened, so a growing number of ingenious and inventive people have been drawn into developing new approaches for humanity to adopt.

Bullock carts, Cambodia. Simple technical improvements, some of which have been known and used for hundreds of years in the developed world, could double or triple the efficiency with which energy is used. Tyres, better yokes for the bullocks, better axles would all improve the efficiency of what is still, for many millions of people, the only means of transport.

Don't waste it

DEPARTMENT OF ENERGY

Energy
In no field is this more true than that of energy. Both its use, and its non-use, help to increase ecological stress. We have seen elsewhere some of the problems that energy use creates: acid rain from burning coal; photochemical smogs from vehicle exhausts; radioactive waste from nuclear power stations and carbon dioxide from the use of any carbon-based fuel. The unavailability of alternatives to wood in many developing countries, or the limitations placed on agricultural productivity by the absence of cheap energy, also have damaging ecological consequences which have been illustrated earlier. In the wake of the oil-price rises of the 1970s, we face a world in which oil and gas are running out, nuclear power has turned out to be uneconomic and unacceptable almost everywhere and the

231

Above: since the first oil crisis in 1973, many governments have launched massive programmes to save energy. The relationship between energy use and economic growth has now been significantly altered, with up to 20 per cent more economic activity being generated for each unit of energy used.

Six million homes are already being heated by the Sun. By the year 2000 this will have risen to over 30 million. Wind, wave and tidal power, as well as fuel from fuel crops, are all indirect forms of solar energy.

use of coal is constrained by the environmental impacts of its extraction and burning. These converging pressures have stimulated investigation of two previously neglected avenues, both of which promise to lower the ecological impact of energy use, while maintaining security of supply.

The first thing we can do is to increase the efficiency with which we use the energy we already have. Cheap energy created very wasteful habits. Now that energy is expensive, ways to get the most out of every bit used are being developed. In Britain, over a third of the energy used to heat homes and water could be saved without any loss of comfort, and with a substantial saving of money. Industrial energy use could be reduced by a fifth and transport energy consumption by over a half. These are modest targets which assume no technological breakthroughs and no further rise in fuel prices. Under more likely assumptions some calculations suggest that the United Kingdom could have three times its present level of economic activity and still use only 40 per cent of the energy it is now using. Such a pattern could be repeated throughout the industrialized world.

The advantages to the environment in terms of reduced pollution and resource use are apparent. What is less obvious are the gains to the economy in increased competitiveness of goods, lower import costs and the creation of large numbers of additional jobs. There are encouraging signs that many governments and industries are beginning to realize the advantages of a more energy-efficient society. A good many governments have now launched programmes to stimulate the insulation of walls and attics in homes as well as giving grants to industry to replace inefficient equipment. Motor manufacturers have reduced the size of cars, improved their traction and aerodynamic performance and developed more economic engines. Cars that previously averaged 20 to 30 miles per gallon now comfortably reach 40 to 60

miles per gallon, and further improvements are well within reach.

However damaging the effects of energy waste in the industrialized world, they are much more so in the developing countries. At first sight, the relative scarcity of energy resources in the poorer countries would suggest their more efficient use. Paradoxically, the opposite is more often true. Scarce and expensive wood or charcoal is often burned in open fires or badly designed stoves which waste most of the heat. Draught animals, on which much of the developing world still relies for motive power, are poorly harnessed to vehicles which rarely have rubber tires, or have poorly designed axles. Simple changes in traditional designs of stoves or in methods of harvesting can cut fuel use by up to a third. Eliminating waste is the first step towards a sustainable society. In the energy sector at least there are hopeful signs that real progress is now being made towards improving efficiency.

The second avenue that is now being explored with growing enthusiasm is the

Odeillo in the French Pyrenees. This solar furnace produces temperatures of up to 3,000°C by focussing sunlight reflected from 60 computer-controlled mirrors.

use of renewable sources of energy. The sun, wind, waves, tides, hot rocks and organic materials so long dismissed as marginal by energy planners are beginning to look increasingly attractive. Until the widespread use of coal and, later, oil, mankind was reliant entirely on renewable sources of energy. The prime source was the sun itself, both directly for heating buildings, often carefully designed to make maximum use of the sun's heat, and indirectly as wind to move vessels or grind corn, as wood for fuel, as water to turn mills. As energy pressures have intensified, so planners have taken a fresh look at some of these old uses as well as developing wholly new technologies. By combining design concepts that have been known for centuries with modern building materials we can now build houses that even in the coldest climates receive 75 to 90 per cent of their energy from the sun. Simple solar collectors are now to be found heating water in over two million Japanese buildings and, within ten years, the Japanese expect a third of their buildings to be equipped in this way. Sixty companies in twenty countries are now producing photovoltaic cells which convert sunlight directly into electricity. Water pumps in Tunisia, electronic calculators in Europe, medical refrigerators in Africa and telecommunications systems in Papua New Guinea are now all powered by photovoltaic electricity. By the middle of the next century as much as a third of the world's electricity could come from the sun.

These options only scratch the surface of what is possible. Brazil supplies nearly twenty per cent of its transport fuel with alcohol derived from sugar cane grown as an energy crop. In China, biogas plants, which convert animal manure into methane, supply the equivalent of 22 million tonnes of coal a year. Twelve per cent of Munich's electricity comes from burning garbage and other wastes. France has identified 90,000 potential sites for small-scale hydroelectricity generators. The Soviet Union plans to deploy 150,000

wind turbines by 1990. In the United Kingdom, experiments are underway to tap the heat of the rocks that lie beneath our feet.

It is now quite clear that there is no shortage of energy, only of the means to make use of it. Throughout the world scientists, energy planners and investors are now beginning to explore seriously the potential of renewable sources of energy.

A motorway interchange in the heart of the British countryside. In Britain, seven million hectares of agricultural land are lost annually to roads and settlements. Much of the land left at this interchange cannot be used by agriculture – or even by wildlife – as it is isolated by roads.

As more research is carried out and more investment made, so new opportunities are being discovered and costs are falling. As the means to make use of renewable energy become available, so the other benefits of their use become apparent. For the most part, their impact on the environment is low; moreover, they are much more equitably distributed than other fuels and start-up capital is often much lower than for orthodox energy technologies. These factors reduce the disadvantages suffered by the poor countries. Yet most governments still regard renewable energy as the Cinderella of energy policy, to be considered, if at all, only as an afterthought. In any state, a substantial shift of financial, intellectual and administrative resources in favour of the renewables would strengthen both the

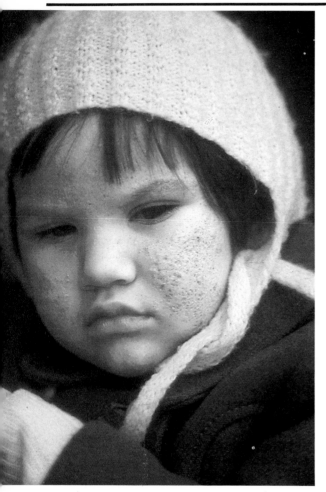

economy, by ensuring adequate supplies of energy, and the environment, by reducing the ecological stresses of energy use.

Land

No sustainable society can be achieved without a positive resolution of the food–land–population equation. But, as we have seen, the margin for manoeuvre is not very wide. Populations will continue to rise for some time, even if we immediately adopt vigorous policies to control their growth. The productivity of our most productive areas cannot be increased as fast in the future as it has in the past. The amount of new land that can be brought into production is balanced by the amount

that is being lost to production through urbanization and soil degradation. Yet this narrow window of opportunity is being constantly widened by human ingenuity.

Agricultural productivity can be increased, and quite dramatically so, in many developing countries. More reliable and better quality irrigation can more than treble yields in the rice paddies of China or the Philippines; mixed cropping of grains and leguminous plants, which fix their own nitrogen, can increase yields without the need for expensive artificial fertilizers. Where inorganic fertilizers or pesticides are used, better understanding of how they work allows much smaller quantities to be used in obtaining higher yields. Some steps, such as the

Seveso, Italy. One of the most publicized environmental tragedies of recent times occurred when an explosion in a chemical plant in Seveso, Italy, spread a few pounds of the deadly poison, dioxin, over the surrounding countryside. As well as inducing birth defects in both humans and animals, dioxin causes severe immediate illness and skin

diseases such as chloracne (left). Because the organo-chlorine compounds, to which dioxin belongs, are persistent, the authorities had no option but to evacuate the town. Ten years later it is still empty.

construction of grain silos in Mauritania to improve storage, can eliminate waste. Breeding disease-resistant animals, reseeding pastures, and increasing the off-take from herds, are all increasing the yield from livestock. Higher yields not only bring the direct benefit of increased food, they also allow farmers to break out of the vicious cycle of reducing fallow periods and constantly expanding the area under cultivation to maintain production.

But increasing productivity cannot alone hope to keep the equation in balance. Unless the bleeding away of land to urbanization, desertification and erosion is staunched, population growth will soon overtake any conceivable growth in productivity. Reserving good croplands

for crops is an essential step towards a sustainable society. Much of the developed world would do well to follow the example of the French *Sociétés d'Amanagement Foncier et d'Etablissment* (SAFERs). These local associations are empowered by law to pre-empt any sale of farmland and, although they only purchase about 12.5 per cent of the farmland put up for sale, they exercise strong influence on the market to discourage non-agricultural development on productive land. Although government decisions have long determined the way land is used, most governments have, until recently, been reluctant to establish tough national guidelines on land use policy. Now an increasing number of countries are taking steps to identify their most productive land and to establish legal and financial barriers to its inappropriate exploitation.

Elsewhere in the world, a growing effort is going into halting the spreading deserts. In Israel, parts of the Negev desert that have been overgrazed and deforested since time immemorial are once again productive as controlled grazing, improved dry-land farming and innovative irrigation transform the desert. In China, Rajasthan in India, and Somalia sand dunes are first stabilized and then planted with hardy trees and grasses. After five years the land is once again ready for grazing. In India programmes are under way to reclaim nine million hectares of saline desert by combining applications of gypsum and fertilizer with the planting of salt-resistant varieties of grain, rice or trees.

Industry

Industry is often considered the environment's major enemy. Its damaging effects, from foul air and dirty water to mountains of mining spoil, are all too obvious. Few people are likely to forget the Seveso accident which contaminated seven hundred square miles of Italy with the deadly poison dioxin, causing disease and birth defects. Dark though industry's reputation is, there is a silver lining. Public understanding, tough legislation and rising costs are causing a growing awareness in industry that one man's pollution is another man's profit. One Union Carbide plant, forced by environmental pressure to take a fresh look at its processes, found that it could save over a million dollars a year, and reduce pollution, just by tightening some flanges. Other examples are more costly, but no less profitable. A French oil company faced with a disposal cost of 2.5 million francs for hydrocarbon wastes instead found a market for the products and, after an investment of eleven million francs, turned in a profit of five million francs a year. The 3M company, one of the most aggressive

innovators in this area, introduced a programme which in nine months eliminated 70,000 tonnes of air pollution and 500 million gallons of waste water, while saving the company over ten million dollars.

Industry is not only criticized for processes which pollute, but also for creating products which are wasteful of scarce resources. The emergence of planned obsolescence and the throwaway container was such a glaring instance of wasted energy and resources as to arouse hostile reactions from consumers. This consumer reaction, coupled with rising energy costs, has stimulated many countries into developing vigorous

Millions of cars are already re-cycled each year. The re-use of ferrous scrap not only saves a tenth of the energy used to produce steel from raw materials, but reduces air pollution by four-fifths and water pollution by three-quarters.

recycling industries. The Japanese economy, very conscious of its dependence on imported oil and minerals, has increased the proportion of waste materials recycled from 16 per cent to 46 per cent. So dependent is the Japanese car industry on recycled materials that it imports them from other countries. East Germany has over 11,000 collecting points for newspapers and glass, and Norway insists that car buyers pay a deposit which can be recovered when the car is turned in at an authorized recovery centre. Some estimates suggest that as much as two–thirds of the material resources we now use could be recycled.

These last examples show just three

Top: one tonne of recycled newsprint saves one tonne of .wood, about twelve trees.

of the routes that must be followed if we are to achieve a sustainable society. Equally potent possibilities exist in other sectors: in forestry, in water conservation and in wildlife protection. Down the whole environmental agenda, a growing awareness of the problems has spurred human inventiveness and ingenuity toward technical innovation and the social transformation which goes hand in hand with it. The result is, as Barbara Ward, one of the greatest of global influences on environmental thinking wrote just before her untimely death:

'For an increasing number of environmental issues the difficulty is not to identify the remedy, because the remedy is now well understood. The problems are rooted in society and the economy – and in the end in the political structure. Foresters know how to plant trees, but not how to devise methods whereby villagers in India, the Andes or the Sahel can manage a plantation for themselves. Biologists know where to draw boundaries for nature reserves, but cannot keep landless peasants from invading them to grow food or cut fuel.

The solutions to such problems are increasingly seen to involve reforms in land tenure and economic strategy, and the involvement of communities in shaping their own lives.'

At the Stockholm Conference, the Chinese delegation said, 'Of all the things in the world, people are the most precious.' This is true. But it is also true to say that they are also the most destructive – of themselves and of every other living being. We know what we must do if we wish to avoid the fate of our predecessor civilizations. We must stabilize our population; we must achieve the sustainable use of the biological resources, and we must eliminate waste from our use of the non-renewable resources. We already know how to do all of these things. What we lack are the personal commitments that translate into the political will to bring our knowledge, our ingenuity, our inventiveness to bear on the crisis in which we find ourselves. We face, as perhaps no other civilization has faced, a choice of direction.

If we allow present trends to continue then we will continue to degrade the ecological base of our civilization. It is unlikely that anything as dramatic will happen to us as happened to the Mayans. We will simply face a future in which the prospects for each succeeding generation get slowly worse, and where those at the margin, the 800 million absolute poor, die rather faster. Civilization will not collapse, just crumble. But we have not yet passed the point of irrecoverability for most of our degraded ecosystems, for the fisheries, croplands and forests on which we depend. If we choose to invest our capital, our minds and our skills on restoring and sustaining these systems, on opening up the several roads to a sustainable society, then the prospect, if not one of the dramatic increases in health we have so far witnessed this century, will be one which offers each succeeding generation the hope that life will get better.

Robert Frost wrote:

Barbara Ward, pioneer of the ecology movement.

'Two roads diverged in a wood, and I –
I took the one less travelled by,
And that has made all the difference.'

The path of wisdom, foresight, cooperation and restraint has always been 'less travelled by'. But it has retained none the less a persistent attraction. If there is one, central message in this book, it is that we have a choice, that the outcome does not depend on the blind forces of Nature or the impersonal workings of the economy. It depends on how each one of us chooses to live our life, how we choose to vote, what we choose to teach our children. A sustainable world in which the economy is in equilibrium with the ecology, in which mankind is in harmony with Nature, is not out of reach. It is within our power to do everything necessary to reach that world, if each one of us so chooses.

Tom Burke

In the past, Man lived in harmony with Nature. When we choose to, we can do so now. If our children and their descendants are to have a future, to live in balance with Nature, in partnership with our ecology, must become the basic rule of our lives.

Index

Page numbers in *italics* indicate the sites of major references.

Mexico, 33, 47, 179, 185; desertification, 24, 67; emigration from, 24; tribal cultures, 163–4, 165, 166; urbanization, 151; water pollution, 117
Mexico, Gulf of, 121
Mexico City, 131, 154, 155
Middle East: palaeoclimates, 185; see also individual countries
millet, see cereals
Minimata Bay, 120
mining, 109
molluscs, 27
monoculture, 50, 61, 164
Montreal, 186
morphine, 95
Moscow, 155
Munich, 234
mussels, 120

Nairobi, 73, 215
Namche Bazaar, 12, 13, 14
Namdu, 9
Namibia, 24
natural gas, 26, 231
Negev-Sinai Desert, 60, 238
Nepal, 7–9, 150, 204, 207–10; deforestation, 7–9, 12–13, 207; National Parks, 13–15
Netherlands, 137, 138, 186
New Caledonia, 104
New Guinea, see Papua New Guinea
New Jersey, 155
New Orleans, 186
New York, 153, 155, 186
New York State, 137
New Zealand, 13–14, 83, 150
Niamey, 80
Nicaragua, 158
nickel, 120, 137
Niger, 213; agriculture, 74; drought in, 70–74 passim; forestry, 80
Nile, 28–9
Nippon Oil, 97
nitrates, 116, 117, 119, 142; see also fertilizers
nitric acid, 117, 135, 142
nitrite poisoning, 119
nitrogen, 144, 179, 200; oxides, 116–17, 131, 135, 145, (nitrous) 138, 145
nitrogen-fixing, 62, 77, 79, 102, 165, 236
noise pollution, 155
Non-Government Organizations (NGOs), 11–12
Norbu, Mingma, 14–15
North America: birds, 33; forests threatened, 135, 139; fossil-fuel consumption, 186–7; palaeoclimates, 185; peoples of, see Amerindians, Eskimos; urbanization, 150; see also individual countries
North Sea, 123
Norway: acid rain in, 134, 137, 138, 142; fishing, 30; recycling policy, 239
Norwegian Sea, 123
Nouakchott, 161
nuclear power, see power generation
nuclear testing, 133
nuclear war, 169, 191, 195

nuclear weapons, 123, 191, 196
nutrition, see diet

Oaxaca Valley, 166
Ob River, 174
oceans: absorption of carbon dioxide, 123, 180; pollution, 120–29
oil (petroleum), 26, 96, 102, 126, 183, 197; and air pollution, 130, 131; in Antarctica, 9; contaminants, 96–7, 133; prices, 30–31, 62, 97, 99, 154, 198, 200, 231; slicks, 190; see also fossil fuels, hydrocarbons, industrial pollution, pollution
Okinawa, 97
Oklahoma, 50
Ontario, 137, 143
OPEC, 31, 62, 97, 183
Osaka, 186
Ottowa, 139
Ouagadougou, 80
overcropping, 60–65; in the Sahel, 70, 218; see also agriculture (intensive)
overgrazing, 29, 60–65, 199; in Ethiopia, 216; in Haiti, 210; in the Negev, 238; in the Sahel, 70, 80, 218
overpopulation, see population growth
oxygen, 144, 179, 200; cycling rate, 182; in water, 109, 118; see also ozone
ozone, 144
ozone layer, 143–6

Pacific Ocean: marine organisms, 27–8; pollution, 120, 122, 129
Pakistan, 185; desertification, 60; population growth, 24; soil erosion, 63–4, 69
Pana dynasty, 207
Panama Canal, 27; proposed new, 26–8
panda, giant, see giant panda
pandemic diseases, 19
Pangboche, 12
Papua New Guinea, 163, 164, 165–6, 234
paraffin, see kerosene
Paris, 155
Park Authority, National (Nepal), 13–14
Partial Nuclear Test Ban Treaty (1963), 133
passenger pigeon, 86
peanut, 62, 80, 166, 218
Peking, 155
Peligre Dam, 212
perch, 137
Peru, 30, 124, 150
pesticides, 31, 46, 96, 236; natural, 165–6; and water pollution, 51, 116, 119, 120–21
pests, crop, 50, 92, 96
petroleum, see oil
pH, definition of, 137
Phacelia, 84
Philadelphia, 133
Philippines, 47, 165, 168, 220–26; deforestation, 220, 226; fisheries, 220; future agriculture, 236; population growth, 226
phosphates, 109, 116, 119; see also fertilizers